Putting the "Make" in Maker

The Italian Trade Agency (ITA) organized this year's participation of 15 Italian makers as part of its ongoing initiatives supporting the country's activities in the global maker movement.

At this year's World Maker Faire in New York, Italy's prominence within the global maker movement was well represented, within the Italian Makers Pavilion, coordinated by the Italian Trade Agency (ITA), Italy's government agency for trade and development. The pavilion featured 15 of Italy's finest makers, many of which participated for the first time ever at a Maker Faire outside of Italy, covering a wide gamut of offerings ranging from fashion to alternative energy, from sensor technologies to the Internet of Things and to 3-D printers. Their intent was to forge new partnerships and promote their products and services with participants and visitors. Check out the Makers who were selected by ITA Rome to participate in the ITA's 2017 World Maker Faire Italian Maker Pavilion at www.makersitalia.com.

Massimo Banzi, Co-founder of Arduino, stops by the ITA's Italian Makers Pavilion.

"We were very excited to be part of Maker Faire 2017, an initiative truly inspired by out of the box thinking. Our esteemed makers reflected in their innovative products and services those changes taking place in the world of manufacturing today", said Marco Saladini, Trade Commissioner of the Italian Trade Agency's Chicago Office. "The basic paradigm is moving to a new approach, where three traditionally distinct phases collapse into one, combining design, production and sale all at the site of the actual retail point. Italian makers are at the forefront of this process and we strongly believe that international exposure will help them to grow and prosper.

The ITA pavilion received upon the conclusion of the show, a Best Pavilion award by Maker Media, Inc.'s editorial board.

We", concludes Saladini, "are definitely aware of the historical significance of this new avant-garde and of its business potential, therefore we will continue to support its efforts to establish new ties with US counterparts."

Italian makers have contributed quite substantially to the global maker movement with such technologies as Arduino, considered by many as the DNA of countless Maker projects. Italy boasts a network of approximately 100 Makerspaces and FabLabs. See their locations at http://www.makeinitaly.foundation/makers/.

Maker Faire Rome, which started less than 10 years after the first maker Faire in the Bay Area (San Francisco) is now embarking on its 6th year, having become the world's third largest faire with over 110,000 visitors and 600 inventions coming from over 65 nations. The 2017 edition will take place on December 1-3, 2017 at the Fiera di Roma in Rome, Italy (http://www.makerfairerome.eu/en/).

The Italian Trade Agency's offices of Rome and Chicago are already planning its next Italian Maker Pavilion at the Bay Area Maker Faire in San Mateo, California, scheduled for May 19-20th 2018. Stay tuned for updates using the contact and reference information on the side and below.

For information about the ITA's activities on innovation and start-up companies
Site: www.innovationitaly.it

Italian Maker Participants:

CONINTECOM SRL
www.guarotti.com

CYBERTEC SERVICES SRL
Progetto Computarte
www.computarte.it

ELEMIZE TECHNOLOGIES SRL
www.elemize.com

EMBIT SRL
www.embit.eu

ENERGY2GO
www.energy2go.it

FABIO ROMOLI
www.horgonic.com

FABLAB CATANIA
www.fablabcatania.eu

FABTOTUM
www.fabtotum.com

FONDAZIONE MONDO DIGITALE
www.mondodigitale.org/it

FONDERIE DIGITALE
Rete di imprese White3 Srl
www.fonderiedigitali.org

JESSICA GASTALDO
www.facebook.com/AMENOstudio

MORPHEOS
www.morpheos.eu

NEXT INDUSTRIES SRL
www.nextind.eu

VANZOTECH SRL/
CRISTIANO VANZLINI
www.vanzotech.com

WEAR SRL
www.wear-mobile.com

For more information on the Agency's US activities related to the Italian maker movement, contact its Chicago Office via the following:

Italian Trade Agency
401 N. Michigan Avenue, Suite 1720
Chicago, Illinois 60611
Tel. 312-670-4360
Email: chicago@ice.it
Site: www.makersitalia.com
Site: www.italtrade.com/usa
Twitter: @ITAChicago
Twitter: @ITAtradeagency

CONTENTS

Make: **Volume 60** December 2017/January 2018

Ultimate Guide to
DESKTOP
FABRICATION
2018

PRINTS ON THE COVER:
Dragon: MakerBot, remixed for
multicolor by Mosaic
Droplet Vase: Vcrettenand
Zortrax Voronoi Sphere: ZRAFT
Moon City design: Jukka Seppänen

Photo: Hep Svadja

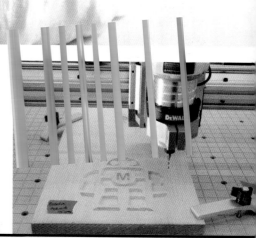

Hep Svadja, MakerBot/Mosaic

Make:

> "If I can print a necklace from home, why not print my clothes from home, too?"
>
> — Danit Peleg, fashion designer

EXECUTIVE CHAIRMAN & CEO
Dale Dougherty
dale@makermedia.com

CFO & PUBLISHER
Todd Sotkiewicz
todd@makermedia.com

VICE PRESIDENT
Sherry Huss
sherry@makermedia.com

EDITORIAL

EXECUTIVE EDITOR
Mike Senese
mike@makermedia.com

SENIOR EDITOR
Caleb Kraft
caleb@makermedia.com

EDITOR
Laurie Barton

MANAGING EDITOR, DIGITAL
Sophia Smith

PRODUCTION MANAGER
Craig Couden

EDITORIAL INTERN
Jordan Ramée

CONTRIBUTING EDITORS
William Gurstelle
Charles Platt
Matt Stultz

CONTRIBUTING WRITERS
Philip J. Angileri, Adam Casto, Jeremy Cook, Matt Dauray, DC Denison, Kelly Egan, Chad Elish, Sean Fairburn, Matt Griffin, Shawn Grimes, Bob Knetzger, Joseph Kowen, Art Krumsee, Seng-Poh Lee, Lisa Martin, Darius McCoy, Charles Mire, Clement Moreau, Evan Morgan, Simon Norridge, Ryan Priore, Jonathan Prozzi, Jen Schachter, Tasker Smith, Joe Spanier, Evan Stanford, Mandy L. Stultz, Sarah Vitak, Terry Wohlers, Chris Yohe

DESIGN, PHOTOGRAPHY & VIDEO

ART DIRECTOR
Juliann Brown

PHOTO EDITOR
Hep Svadja

SENIOR VIDEO PRODUCER
Tyler Winegarner

MAKEZINE.COM

TECHNICAL PROJECT MANAGER
Jazmine Livingston

WEB/PRODUCT DEVELOPMENT
David Beauchamp
Bill Olson
Sarah Struck
Alicia Williams

CONTRIBUTING ARTISTS
Bob Knetzger, Rob Nance

ONLINE CONTRIBUTORS
Halil Aksu, Cabe Atwell, Ivy Bardaglio, Gareth Branwyn, Chiara Cecchini, Jon Christian, Wolfram Donat, Chad Etzel, Liam Grace-Flood, Keith Hamas, Jess Hobbs, Tom Igoe, Zach Kaplan, Justin Klein Keane, Christoph Klemke, Chloe Kow, Becky LeBret, Joel Leonard, Matt Lorenz, Jayson Margalus, Goli Mohammadi, Pete Prodoehl, Sean Michael Ragan, Bridget Rigby, Toshinao Ruike, Gordon Styles, Christine Sunu, Leah Swinson, Andrew Terranova, AnnMarie Thomas, Phillip Torrone, Ben Vagle, Tom Van de Wiele, Shawn Van Every, Sarah Vitak, Glen Whitney, Yan Zhang

PARTNERSHIPS & ADVERTISING
makermedia.com/contact-sales or partnerships@makezine.com

DIRECTOR OF PARTNERSHIPS & PROGRAMS
Katie D. Kunde

STRATEGIC PARTNERSHIPS
Cecily Benzon
Brigitte Mullin

DIRECTOR OF MEDIA OPERATIONS
Mara Lincoln

BOOKS

PUBLISHER
Roger Stewart

EDITOR
Patrick Di Justo

PUBLICIST
Gretchen Giles

MAKER SHARE

DIRECTOR, ONLINE OPS
Clair Whitmer

CONTENT & COMMUNITY MANAGER
Matthew A. Dalton

LEARNING EDITOR
Keith Hammond

DESIGN INTERN
Pravisti Shrestha

MAKER FAIRE

EXECUTIVE PRODUCER
Louise Glasgow

PROGRAM DIRECTOR
Sabrina Merlo

MARKETING & PR
Bridgette Vanderlaan

COMMERCE

PRODUCTION AND LOGISTICS MANAGER
Rob Bullington

PUBLISHED BY

MAKER MEDIA, INC.
Dale Dougherty

Copyright © 2017 Maker Media, Inc. All rights reserved. Reproduction without permission is prohibited. Printed in the USA by Schumann Printers, Inc.

Comments may be sent to:
editor@makermedia.com

Visit us online:
makezine.com

Follow us:
🐦 @make @makerfaire @makershed
google.com/+make
makemagazine
makemagazine
makemagazine
twitch.tv/make
makemagazine

Manage your account online, including change of address:
makezine.com/account
866-289-8847 toll-free in U.S. and Canada
818-487-2037,
5 a.m.–5 p.m., PST
cs@readerservices.makezine.com

CONTRIBUTORS

What's the next big thing in Digital Fabrication?

Jen Schachter
Baltimore, MD
(Asteroid review)

Consumer-level subtractive manufacturing systems are catching up to additive! Streamlined hardware, approachable CAD/CAM, and accessible to hobbyists.

Matt Dauray
Pawtucket, RI
(Dremel 3D45 review)

Bot farm-style production. With reliable printers, magnetic flexible beds, and affordable robot arms, I'm excited to see makers design fully automated production.

Philip J. Angileri
Pawtucket, RI
(Raise3D N2 review)

I'm really excited to see desktop 3D printers moving toward closed loop systems. The future of these machines is in their ability to sense the outside world and react to it.

Shawn Grimes
Baltimore, MD
(It's a Wrap)

Developments in circuit board fabrication that make electronics prototyping more accessible, like milling and pick-and-place. Exciting to create small runs of hardware projects.

Issue No. 60, December 2017/January 2018. Make: (ISSN 1556-2336) is published six times a year by Maker Media, Inc. in January, March, May, July, September, and November. Maker Media is located at 1700 Montgomery Street, Suite 240, San Francisco, CA 94111. SUBSCRIPTIONS: Send all subscription requests to Make:, P.O. Box 17046, North Hollywood, CA 91615-9588 or subscribe online at makezine.com/offer or via phone at (866) 289-8847 (U.S. and Canada); all other countries call (818) 487-2037. Subscriptions are available for $34.95 for six issues in the United States; in Canada: $39.99 USD; all other countries: $50.09 USD. Periodicals Postage Paid at San Francisco, CA, and at additional mailing offices. POSTMASTER: Send address changes to Make:, P.O. Box 17046, North Hollywood, CA 91615-9588. Canada Post Publications Mail Agreement Number 41129568. CANADA POSTMASTER: Send address changes to: Maker Media, PO Box 456, Niagara Falls, ON L2E 6V2

STATEMENT OF OWNERSHIP, MANAGEMENT AND CIRCULATION (required by Act of August 12, 1970: Section 3685, Title 39, United States Code). 1. MAKE Magazine 2. (ISSN: 1556-2336) 3. Filing date: 10/1/2017. 4. Issue frequency: bimonthly. 5. Number of issues published annually: 6. 6. The annual subscription price is $34.95. 7. Complete mailing address of known office of publication: Maker Media, Inc. 1005 Gravenstein Highway North, Sebastopol, CA 95472. Contact person: Kolin Rankin. Telephone: 305-441-7155 ext. 225 8. Complete mailing address of headquarters or general business office of publisher: Maker Media, Inc. 1700 Montgomery Street; Suite 240, San Francisco, CA 94111. 9. Full names and complete mailing addresses of publisher, editor, and managing editor. Publisher, Todd Sotkiewicz, Maker Media, Inc., 1700 Montgomery Street; Suite 240, Editor, Mike Senese, Maker Media, Inc., 1700 Montgomery Street; Suite 240, San Francisco, CA 94111, Managing Editor, N/A, Maker Media, Inc., 1700 Montgomery Street; Suite 240, San Francisco, CA 94111. 10. Owner: Maker Media, Inc.; 1700 Montgomery Street; Suite 240, San Francisco, CA 94111. 11. Known bondholders, mortgages, and other security holders owning or holding 1 percent of more of total amount of bonds, mortgages or other securities: None. 12. Tax status: Has Not Changed During Preceding 12 Months. 13. Publisher title: MAKE Magazine. 14. Issue date for circulation data below: Oct/Nov 2017. 15. The extent and nature of circulation: A. Total number of copies printed (Net press run). Average number of copies each issue during preceding 12 months: 119,913. Actual number of copies of single issue published nearest to filing date: 119,979. B. Paid circulation. 1. Mailed outside-county paid subscriptions. Average number of copies each issue during preceding 12 months 58,170. Actual number of copies of single issue published nearest to filing date: 57,494. 2. Mailed in-county paid subscriptions. Average number of copies each issue during the preceding 12 months: 0. Actual number of copies of single issue published nearest to filing date: 14,730. 4. Paid distribution through other classes mailed through the USPS. Average number of copies each issue during the preceding 12 months: 0. Actual number of copies of single issue published nearest to filing date: 0. C. Total paid distribution. Average number of copies each issue during preceding 12 months: 73,092. Actual number of copies of single issue published nearest to filing date: 72,224. D. Free or nominal rate distribution (by mail and outside mail). 1. Free or nominal Outside-County. Average number of copies each issue during the preceding 12 months: 708. Number of copies of single issue published nearest to filing date: 753. 2. Free or nominal rate in-county copies. Average number of copies each issue during the preceding 12 months: 0. Number of copies of single issue published nearest to filing date: 0. Free or nominal rate copies mailed at other Classes through the USPS. Average number of copies each issue during preceding 12 months: 2,932 . Number of copies of single issue published nearest to filing date: 3,532 . 3. Total free or nominal rate distribution outside the mail. Average number of copies each issue during preceding 12 months: 3,639. Actual number of copies of single issue published nearest to filing date: 4,285. F. Total free distribution sum of 15c and 15e. Average number of copies each issue during preceding 12 months: 76,730. Actual number of copies of single issue published nearest to filing date: 76,510. G. Copies not Distributed. Average number of copies each issue during preceding 12 months: 43,183. Actual number of copies of single issue published nearest to filing date: 10,469. H. Total (sum of 15f and 15g). Average number of copies each issue during preceding 12 months: 95.26% Actual number of copies of single issue published nearest to filing date: 116,979. I. Percent paid. Average percent of copies paid for the preceding 12 months: 94.0%. 16. Electronic Copy Circulation. Average number of copies each issue during the preceding 12 months: 20,619. Actual number of copies of single issue published nearest to filing date: 21,685. B. Total Paid Print Copies (Line 15c + Paid Electronic Copies (Line 16a). Average number of copies each issue during preceding 12 months: 93,100. Actual number of copies of single issue published nearest to filing date: 98,195. C. Total Print Distribution (Line 15f) + Paid Electronic Copies (Line 16a). Average number of copies each issue during preceding 12 months: 97,339. Actual number of copies of single issue published nearest to filing date: 95.64%. I certify that 50% of all distributed copies (electronic and print) are paid above nominal price. Yes. Report circulation on PS Form 3526 X Worksheet. 17. Publication of statement of ownership will be printed in the Dec/Jan 2018 issue of the publication. 18. Signature and title of editor, publisher, business manager, or owner: Todd Sotkiewicz, Business Manager, I certify that all information furnished on this form is true and complete. I understand that anyone who furnishes false or misleading information on this form or who omit material or information requested on the form may be subject to criminal sanction and civil actions.

PRINTED WITH SOY INK

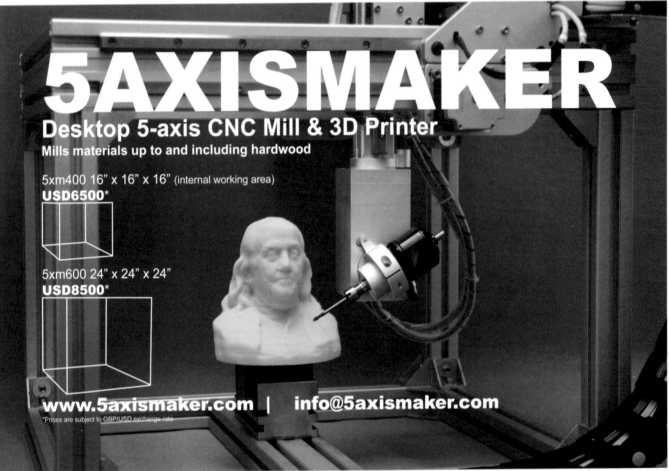

Cosplay Inspiration, Slinging the Perfect Pie, and Material Conflicts

Ken Javes, Keith Hamas

COSPLAY FOR THE WIN!

My two-year-old daughter Alice's request to be a giant robot for Halloween had me scratching my head until I gained some inspiration and education from your cosplay article in Vol. 59 ("Clever Couture," page 16). I managed to put this one together over the past few weeks and she took home first prize at a local costume contest.

–Ken Javes, via email

PERFECTING PIES

At last year's World Maker Faire we gave away a "One-Day Wood-Fired Pizza Oven" (Vol. 53, page 34, makezine.com/projects/quickly-construct-wood-fired-pizza-oven). Keith Hamas was the lucky winner and recently wrote about his experience building the oven and striving to cook the perfect pizza:

"For anyone who dares to try making it, pizza is more than food. It's a time-honored quest that promises challenge and frustration and, most importantly, a sense of delight that transcends bodily nourishment. Building my own outdoor brick pizza oven was a fun and fitting way to start a journey toward making the perfect pie."

Read more at makezine.com/go/slinging-pies

STEALTH ADVERTISEMENTS?

I'm writing about an article that recently appeared on your site, ("DIY Hove Plant," Volume 59, page 56, makezine.com/projects/build-your-own-magnetic-levitating-plant).

Cool idea, nice article, but it's unfortunate (for an article targeted at *makers*) that the only source provided for the key part is the author himself, at a price somewhere between 2 and 3 times the going rate on AliExpress. In the spirit of the movement, it seems reasonable to expect a link to the OEM, rather than exclusively to the author's shop, with no clear way to identify the true original source. Otherwise, it's really just an advertisement and should be labeled as such — not that I mind people trying to make a buck! –*Mike Coffey, via email*

Executive Editor Mike Senese responds:
We quizzed the author Jeff Olson about this question specifically when we were editing the project, and he assured us the levitation unit he's selling is better than the others with same specs found online. But I admit we're pretty much at his mercy on that question, without buying and testing them all.

This is something the editorial team discusses a lot — we're aware of the awkwardness of having a person write a project around a product they sell. But we've also watched and supported communities that have grown as they've contributed *Make:* content — Arduino and Raspberry Pi are two of the more prominent examples. While it's not a perfect apples-to-apples comparison, we do think there's value to supporting aspiring maker pros and their DIY projects.

PRINT SIZE IS TOO SMALL

I've read articles on your website for a while now and I recently decided to subscribe to your print publication. I've got to ask, does anyone actually proof the magazine in print form before it gets published? The font size you use is too small, and it's difficult to read! –*Jeff B, via email*

Art Director Juliann Brown responds:

We'll work on it. ●

MADE ON EARTH

Backyard builds from around the globe

Know a project that would be perfect for Made on Earth?
Let us know: *makezine.com/contribute*

MAKING WAVES

ERINSTBLAINE.COM

For the past two years a very unusual scene has played out in Greensboro, North Carolina. Almost 400 merfolk migrate from all around the world to show off their tails, splash, and play together at North Carolina Mermania. **Erin St. Blaine**'s Mermaid Glimmer stands out.

This gorgeous LED-laden, waterproof, and swimmable mermaid tail weighs 15 pounds and took 3 years to make. Even without the technology the tail shows an amazing level of craftsmanship, constructed with neoprene and a monofin swimming fin, then painted with a scaly texture and accented with rhinestones, gems, and lace.

This was St. Blaine's first large scale LED costuming project. It has 180 addressable waterproof LEDs that she cast in resin or sealed in silicone for additional waterproofing. They're controlled by an Arduino Micro connected via Bluetooth to a custom-made app. This was one of the more challenging parts of the project, since it was St. Blaine's first Android project, and with every redesign of the tail's LEDs, she had to completely overhaul the app. St. Blaine was able to use the FastLED Arduino library to get some amazing different moods and themes working very quickly. She synchronized the lights to a song that she and her husband recorded together.

She says that the project has been really hard and that "the sheer amount of tinkering and tearing down and rebuilding and fixing was unbelievable. If I'd known how hard this would be at the outset, I probably would have given up before I started, but sometimes ignorance is a blessing. When a project is driven by passion, and you refuse to give up, all the frustration can become worth it in the end." —*Sarah Vitak*

Chris Crumley, Robert Minnick

COVERT CREATURES

THOMASDAMBO.COM

For the last 25 years, Copenhagen-based artist **Thomas Dambo** has become an expert at working only with recycled materials. "I've done this since building tree houses as a kid — I would go scavenge with a shopping cart." Dambo, along with his team of two assistants and three interns, source scrap materials from local businesses. They tend to use a lot of pallets, especially on projects outside of Denmark, as they're easy to come by.

His latest builds, *The 6 Forgotten Giants*, are located around the world, from Australia to Germany, Florida, and Puerto Rico. Dambo treats it as a treasure hunt, to "give some mystique and adventure" to the experience.

"A lot of people have forgotten to be curious and explore," Dambo says. "Almost nobody knows what's hiding in their own city. By putting the sculptures in places people don't normally go, I give them both the experience of the sculpture but also the nature. I believe this gives them a much bigger experience than if the sculpture was in the middle of the city. You put effort into finding them." Each of the sculptures is accompanied by a poem engraved on a nearby stone, which gives hints about where to find the others.

Dambo chooses a location, and then lets his materials inspire the type of creature he makes. He brings them to life by making them part of the surrounding environment. They grab trees, lean their back against the hill, and make themselves at home. If he's working locally, Dambo will construct the more detailed parts like faces, hands, and feet at his workshop, and then drive it to the installation site. It takes him about 2 weeks to make a sculpture with the help of 5–10 others.

In addition to his creatures, Dambo runs a small public school out of his workshop for people to come and build, play, and prototype with his recycled materials. "It has become my work and mission to teach others. I believe we need to take better care of our planet and that being better at recycling is a big part of this. I make big, positive, fun, and interactive projects to show people that recycling can be much more than trash."

—*Sophia Smith*

Thomas Dambo

BEE-DIMENSIONAL PRINTER JENNIFERBERRY.DESIGN

As light passes through the illuminated beeswax sculptures of artist and biologist **Jennifer Berry**, it highlights the honeycomb structure. Layers of the delicate material create geometric forms and the natural pattern of the comb is revealed in the shadows. These sculptures are the result of a particularly ingenious 3D printer Berry has developed, the B-Code biological 3D printer, which allows her to collaborate with some unlikely artists: bees!

Berry was a practicing beekeeper, and so was inspired to work with them during her artist's residency at Autodesk in the summer of 2013. She relocated a colony of bees, and gave them a piece of extra

honeycomb to see what they'd make of it. A few days later, she discovered that "the bees had fused the small squares of comb to one another, repaired and smoothed all the damaged edges, and had started rebuilding on the comb. It was beautiful," she recounts.

To better collaborate with the bees, Berry created the B-Code 3D printer. The B-Code is a plastic enclosure that encourages the bees to build sculptural forms. Vents can be added and removed to create passive environmental control. Berry makes a compelling case for calling it a 3D printer: "If you can imagine each bee as an individual print head with its own set of code, determined over time by evolution, and the

internal space within the plastic form as our print bed, then you can get a sense of the B-Code as a biological 3D printer."

Bees won't build directly on the plastic enclosure, so Berry places forms coated in beeswax or arrangements of cut honeycomb, and the bees grow the sculpture from that. Bees are very predictable. They will always build 4–6mm hexagons, and will turn loose honeycomb into something functionally useful for the hive. When the sculpture is somewhere between being visually appealing to humans and useful for the bees, Berry will remove the comb and prepare it for display.

—Lisa Martin

WOODCRAFT®

EST 1928

QUALITY WOODWORKING TOOLS • SUPPLIES • ADVICE®

PERSONALIZED PRECISION FOR YOUR NEXT PROJECTS

With the CNC Piranha® XL , you can accurately create smaller-scale projects like carvings, plaques, ornamental boxes and precision parts machined from wood, soft metals, or plastics — all at a fraction of the price of a full-sized CNC. Create intricate 3-D carving inlays and engravings with its 3-D model library. The CNC Piranha's compact 12" x 24" table is convenient and can easily fit on one end of your bench. The creative potential is unlimited.

162342   CNC Piranha XL®

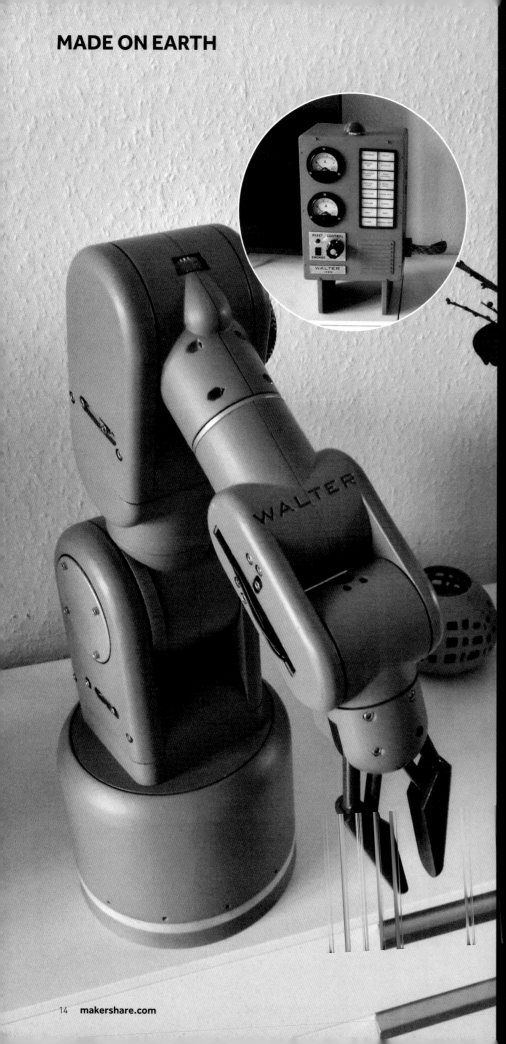

RETRO REPLICA

GITHUB.COM/JOCHENALT

Walter is a 6-axis robot who bears the distinctive '50s style of the former German Democratic Republic. He is the creation of German software architect **Jochen Alt**, who produced a rather sad video of Walter's life as a retired robot arm (youtu.be/XK3WcrrcC8U).

Alt "wanted to build a robot that looks like [him] — vintage." Servos and belt drives control the 6 axes. The gripper is actuated by a servo as well. Alt 3D printed the body, prepped the surface with primer and filler, then painted in a Reseda green that was a standard for factory machinery in times past.

The beauty of this device isn't just skin-deep — it extends to the mechanical design and excellent documentation. Walter has 80 bearings, including one with a diameter of 110mm in the base.

Many robot makers stop as soon as their creation is able to move. Making it move well is part of the process though, and Alt went through some rather involved Bézier curve equations in order to control the arm's trajectory, and set up PC software for trajectory planning as well as hardware in the arm's control cabinet. Data is displayed in OpenGL on the PC and then transmitted to the cabinet for execution.

It took Alt around 30 weeks to build Walter. He worked at night, on the weekends, and two "vacations on the beach when [he] had a lot of time to work on the CAD design."

If you want to build your own Walter, Alt has provided excellent documentation of the project. CAD files, code, and data sheets are on his GitHub (github.com/jochenalt/Walter). —*Jeremy Cook*

Jochen Alt

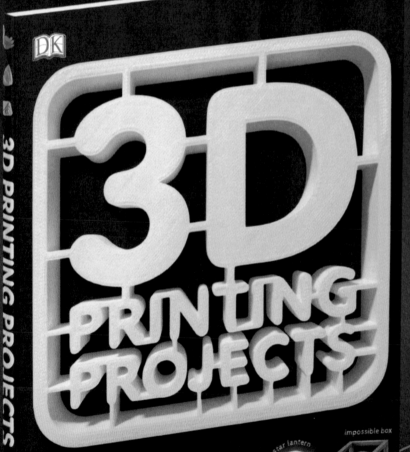

MAKE THEIR HOLIDAYS BRIGHT

3D PRINTING PROJECTS

AMAZING IDEAS TO DESIGN AND MAKE

tablet stand · coat hooks · star lantern · impossible box

SMITHSONIAN

MAKER LAB

28 SUPER COOL PROJECTS

Build · Invent · Create · Discover

USING SCRATCH

STAR WARS CODING PROJECTS

A STEP-BY-STEP VISUAL GUIDE TO CODING YOUR OWN ANIMATIONS, GAMES, SIMULATIONS, AND MORE!

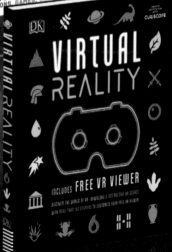

VIRTUAL REALITY

INCLUDES FREE VR VIEWER

DISCOVER THE WORLD OF VR, PROVIDING A DISTRACTIVE VR SCENES WITH MORE THAN 150 STICKERS TO CUSTOMIZE YOUR FREE VR VIEWER

Disney · LUCASFILM

A WORLD OF IDEAS:
SEE ALL THERE IS TO KNOW

www.dk.com

A Fab Family

Founder of 1st Maker Battalion, Inc. and inventor of the Gaming Throne, **SEAN FAIRBURN** is an Emmy award-winning cinematographer, Marine combat veteran, father of five, and inventor-designer dad.

Building a big project together can give long-lasting rewards
Written by Sean Fairburn

My sons and I started building our advanced desk and electronics platform we call Gaming Thrones about four years ago, when my oldest son Caleb was 17, Joshua was 15, and Nathanael was 12. The reason for building the thrones in the first place was to give me and my sons an amazingly comfortable, specifically designed collaboration workspace. Caleb designed the thrones and all their parts in Fusion 360; we hired cabinet shops with CNC routers to cut out the 89 separate pieces for each throne. Getting the design accurate was time consuming, and tooling all the parts with the usual modifications and changes was very difficult and really slowed down progress.

We decided to cut out templates from the parts for future thrones on a table router. That method worked great for two thrones, but each time we wanted to modify the design we'd need a new template for each part. It also increased the time it took to cut all the parts out from 2.2 hours on a CNC to 4 days as every hole, cut, countersink, and roundover had to be done by hand.

LEARNING THROUGH DESIGN
At that point we decided we would need to understand and implement every aspect of the process to give us the flexibility needed to move forward. The main reason for doing this was to create a catalyst for the education of my kids, and give them a practical reason to learn what it takes to move through every aspect of a project and solve and understand each interrelated discipline:

» **Designing** the thrones in Fusion 360 allowed them to learn Fusion and 3D design
» **Engineering** the parts and following the design accurately brought in practical woodworking skills
» **Finishing** each part with sanding and painting created a sense of quality and craftsmanship
» **Assembly** put all the parts together and each aspect of each layer became real and appreciated.

The next step was to look at the cost and ability to manufacture products quickly and efficiently. If we got an order from a company for 12 thrones it would take three months to fulfill by hand. It became obvious that we needed to make our own CNC machine.

MASSIVE ASSEMBLY
Together we set out to make something that would be large, stable, and provide a 4'×8' cutting area. I also wanted to have it take up less floor space and manage dust collection better than conventional flat tables. We decided to make it very strong, with the ability to cut soft metal and hardwoods with ease. We researched each part, and if it could be purchased, like Nema 34 motors and mounts, we would purchase them. We designed the brackets, gantry, and table base to be made from

Welding a very flat, tight, level bed was arguably the toughest part of the entire project.

I encourage testing and bold experimentation, especially with designs and finishes; failure is a mighty teacher.

3"×3"×³⁄₁₆" steel square tube. This opened up a whole new skill to be acquired: welding, including understanding the thermodynamics that make accuracy difficult as welds expand and contract while holding steel parts.

Welding a flat, tight, level bed was arguably the toughest part of the entire project. The center ended up about ¼" higher than the corners over the 5'×10' bed. I expected it to be worse.

Next up was the purchase of linear rail and blocks, which we got from Motion Constrained. Then the rack for the motor pulleys, which required a lot of patience as we had to, again, be very precise and deliberate in setting the pieces then drilling and tapping all the screw holes.

We designed and welded the gantry using parallel tubes to make one piece for strength, and for the flexibility to mount multiple spindles and tools. For this 88"-long steel gantry, we had to affix vertical lifts to the table rail blocks, then weld them precisely into place with no stress for a perfect fit. We maintained exact distance and parallel positioning by clamping wooden blocks between both ends 6¹⁄₁₆" apart, then tack-welding them very quickly to keep them from pulling or twisting apart.

We also designed a custom z carriage to let us mount multiple tools, bolting them in at various heights so that the z travel lead screw would not need to be so long. This allowed us to precisely tram the tools to 90° of the table.

We decided to slant the table 65° so that the dust collection would be easier as gravity would pull chips down into the bottom where the collection port would be. This design also let us easily get to anywhere on a 4'×8' sheet of material, as well as take up less space in the garage.

We installed the motors, power supplies, controllers and wiring, mapped out power, and assembled a custom 220V breaker box to run everything including the 4kW spindle and VFD (variable frequency device), which is like an industrial size router with 6HP of torque.

Caleb decided to gut an old Mac G5 tower and use it as our controller box because it looked cool and you never see Mac stuff around CNC machines. We thought it would be quite funny for people to look at our rig and wonder if it all runs on an old Mac G5.

Soldering became Nathanael's job, putting male and female 4 pin connectors onto each cable for the motors for quick disconnect throughout the system. Measuring and testing became another important skill we learned as all the wires were checked for continuity and properly set to length to keep things tidy.

After all the parts were in place, we needed to do a distance calibration for all three axes, so that 12" in the Mach3 software is 12" actually cut. This is done in a motor tuning section of the program — you set the steps per inch for each axis' motor, compare the actual distance traveled, and then make adjustments to get finer and finer accuracy. We used an X-Acto knife in a drill chuck to mark the distance traveled, until we were repeatedly getting to 0.002" at 96" distance.

MARKED IMPACT
Caleb, now 21, freelances as a designer for hire with a 3D printing company in New Orleans. He previously worked at a high-end cabinet shop, running their CNC, before returning to build our home CNC and work with me on the thrones, and the custom shop stools and tables we sell. When he was younger I would punish him by making him watch Photoshop tutorials. Now he goes right to the web to watch tutorials on anything he wants to learn to do.

Joshua now works at the same cabinet shop making and installing cabinets and building door faces and drawer fronts. When he is not at work he is a wood turner and a very hands-on maker. He's a huge fan of Jimmy DiResta. Joshua has no fear and can make anything.

Nathanael worked in finishing for a while, learning to sand and spray-finish cabinets. He loves this, and has good patience for it. He is also into wood carving and inlay work, as well as calligraphy. He learned Fusion 360 from his older brother Caleb, and teaches it to his younger brother Isaac instead of playing video games.

Even my 11-year-old, Isaac, has now designed a 10-1 scale 2×2 Lego brick in Fusion 360. He knows how to CAM the pieces, tool the parts, import the file to Mach3, load the bits, and cut the wood. He knows exactly what materials to use and how much time it takes to make each Lego. He is now gathering wood for making Christmas presents to sell to others, putting into practice everything he has learned. Makes me very proud.

MY REFLECTIONS
Nowadays, I just get to be Dad, encouraging, supporting, and always watching over. Sometimes I'll get things started then hand them off to the boys. I am retired, but being a cinematographer for 25 years and a U.S. Marine combat vet, I encourage teamwork, good communication, and totally owning the information on the work you are doing — you gotta know your stuff. I stress for my boys to be in the moment and pay attention to the tools. I encourage testing and bold experimentation, especially with designs and finishes; failure is a mighty teacher.

Each project lets me hand more and more to my kids so that they ultimately handle each and every aspect and have an understanding of the total time and materials needed when bidding on a job or selling a product they've made. I am extremely proud of my kids — the time invested in them to become better makers, inventors, and entrepreneurs has unlocked a wide range of capabilities in their future. ◉

NOTE: For helping lift the z carriage along the z-axis, many had asked about pulleys and counterweights, but after testing I found these were not needed. 1,842oz in a Nema 34 motor with a 3-1 reducer gives 345lb of lifting power at 1", so these have no problem lifting the z carriage.

Our Home CNC's Technical Specs:

» 5'×10' bed for 4'×8' cutting area (2" lagniappe)

» 88" gantry width with 3' of floor space used with a 65º bed angle.

» The y-axis can reach 512ipm

» The x-axis can reach 410ipm

» The z-axis can reach 150ipm (lead screw: Igus Drylin, with energy chain carrier)

» Spindle rotates from 8,000RPM to 18,000RPM

The Fairburn Family

FUNCTION OVER FORM

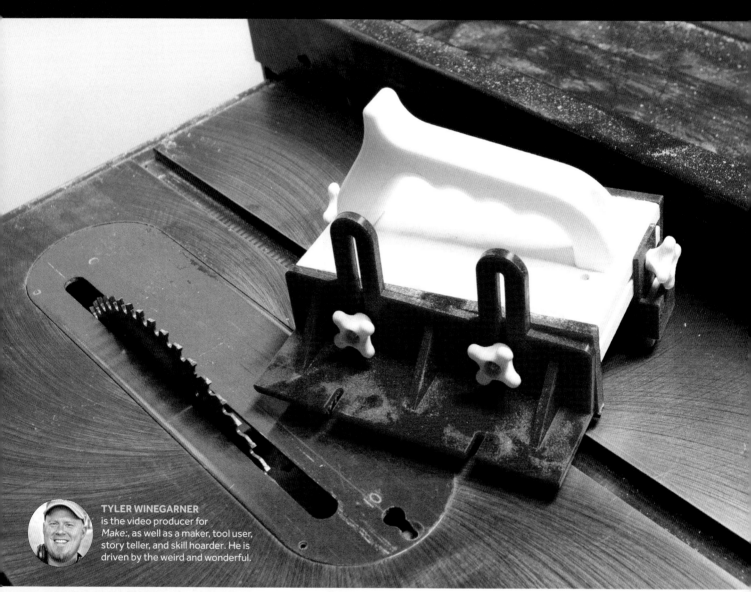

TYLER WINEGARNER
is the video producer for
Make:, as well as a maker, tool user,
story teller, and skill hoarder. He is
driven by the weird and wonderful.

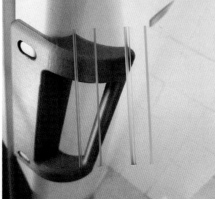

What to do with your printer when you're tired of making Yoda busts and fidget spinners
Written by Tyler Winegarner

3D printing has been accessible to the hobbyist for quite a few years now, and in that time it has earned itself a reputation for being a bit of a toy — printing plastic busts of pop culture icons, Christmas ornaments, trinkets, and silly doodads. But the promise of a 3D printer is so much more than that. You can easily prototype and fabricate simple parts, tools, and jigs, either from your own designs or from online repositories.

Thingiverse and the popular subreddit **/r/functionalprint** are great places to get started putting your printer to practical use — you'll be able to get your head around the sorts of problems that other 3D printing aficionados have solved, and the ways that they solved them. Conversely, you can just wait for something around your house to break. Sooner or later that oven knob will snap off or you'll lose the battery cover to one of your remotes, and you'll have an excuse to start designing or searching for a printable replacement file. (Tip: Search first; there's a good chance that someone else has already designed the part you need.)

Designing your own parts is the way to really unlock the potential of a 3D printer. There are a handful of easily accessible design tools such as SketchUp and Tinkercad, but your time might be better invested in learning some of the more powerful CAD packages like Onshape or Fusion 360. Having some precision measuring tools, like a set of

digital calipers, even cheap ones, will dramatically help with your designing process. You're going to be designing parts that fit with other parts, and you'll need precise measurements. Printing quick test probes to make sure your design fits is also a great practice; it helps to know that your parts are going to fit together correctly before you commit to a long print job, wasting time and material.

ADDING HARDWARE
Having a 3D printer might mean that you have a tiny factory on your desk, but it helps to have some extra bits on hand to help your printed parts work better together. Machine screws and nuts can make the process of joining two pieces together a snap, especially if they require occasional disassembly. Heat-set threaded inserts are a great way to add internal threads to your model to receive a machine screw, or you can model a hexagonal hole in your design to trap a hex nut. It's a good idea to stick to a common screw diameter to simplify your design process, but make sure to get them in a variety of lengths to suit all of your applications. Bearings also make great additions to rotating parts, as well as neodymium magnets to attach your part to metal.

MATERIAL CONCERNS
3D printed parts will almost universally be weaker than parts made through any other fabrication process, but that doesn't mean that 3D printing can't get the job

done. Consumer FFF printed parts won't hold up to high mechanical stresses, but there are plenty of design choices you can make to stack the deck in your favor. Parts will be weakest along their z-axis — the printing layers can separate or shear with enough force. If that's where the stress will be on your part, design it to lie flat on the print bed to offset this aspect. Chamfers and fillets can also make a part far more rigid, as can different infill patterns and densities. Your printing material is worth considering as well — ABS tends to shrink as it cools, but is more flexible and less brittle than PLA. If your printer can reach higher temperatures, you can use stronger exotics like nylon, PETG, and polycarbonate. Then again, the great aspect of 3D printing is the ease of replacement — you can just keep printing spares as you need them.

Once you get the sense for how to design and print parts that solve problems, however minor, you'll likely soon find that your view of the world has transformed — when you find yourself wishing that a specific part existed to solve your unique problem, you'll find yourself with the capability of creating your own solution. Whether you need a very specific mounting bracket, a jig to help you get more use out of a power tool, or maybe just some mounting hooks to help you keep your home or workshop more organized, 3D printing creates a huge opportunity space for you to design, create, and invent. ⊘

Tyler Winegarner, Richard Tran, Peter Pokojny, Gian-Luca Mateo

PRACTICALLY PERFECT

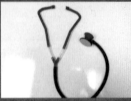

» 3D Printed Stethoscope
Doctors in disadvantaged or war-torn locations are printing their own implements; this stethoscope model costs just a couple of dollars in materials.
thingiverse.com/thing:1182797

» Cord-Cutting Antenna
Our airwaves abound with digital, high-definition television broadcasts, free for anyone with an antenna that can grab the signal. This fractal-based model is an effective design.
thingiverse.com/thing:2471219

» Bike Light Replacement Handlebar Mount
My old headlight mount didn't fit around my new bike's fatter handlebar in a way that I was happy with, so I quickly modeled this one in Fusion 360.
thingiverse.com/thing:2529400

One 3D printing enthusiast is inspiring his peers through entrepreneurship

DARIUS McCOY is the 3D printing manager at the Digital Harbor Foundation where he founded 3D Assistance, a 3D printer repair service for educators, and 3D Printshop, a youth-run printing service.

THE KIDS ARE ALRIGHT

Written by Darius McCoy

My interest in 3D printing started when I was an after-school student at the Digital Harbor Foundation (DHF), a Baltimore-based nonprofit that has a youth- and educator-focused makerspace dedicated to fostering learning, creativity, productivity, and community. My interests there quickly gravitated towards 3D printing, which led me to develop personal business opportunities and help promote youth entrepreneurship.

I remember seeing an iPhone case being printed; I was curious about the whole process. My friend and I both had the same idea of making phone cases and selling them at school. That sparked my first venture into entrepreneurship, a business we called Frozen Lava.

We had a lot of help starting Frozen Lava, but our biggest challenge was actually coming up with a stable product. We decided to make iPhone cases, despite not owning one ourselves, because of their ubiquitous popularity. We wanted to express our Maryland roots by adding designs with crabs or the state outline. In the end, we had a good product but we struggled to sell it. People often said $10 was too much for a case or they wanted something more protective like an OtterBox. We tried our best to maintain the business, but scheduling conflicts made it difficult, and our company eventually fell apart. Although Frozen Lava wasn't successful in terms of sales, I learned a lot throughout the process and was even able to present the project at the White House Maker Faire in 2014 at age 16.

Frozen Lava led to other opportunities. I got a job with the DHF my senior year of high school and started 3D Assistance (3DA), a 3D printer support and repair shop. It's been about two years since 3DA launched and it's still running strong with new employees — the original staff has moved on to college or other jobs like working for a cyber security startup, for example. I'm

especially proud of helping the team develop and grow their professional communication skills, such as learning how to use Slack and email effectively.

At DHF, we were able to capitalize on the success of 3DA to launch 3D Printshop, a small-batch manufacturing operation. 3D Printshop works in collaboration with the University of Maryland and is funded by a grant from the National Science Foundation. We hired a handful of our community's youth to work at 3D Printshop and have taught them how to work with high-quality printers like Ultimakers and Prusas. We're trying to prepare them for real jobs by teaching common workforce skills such as communication, critical thinking, project management, and self-management. The majority of our young staff are now considering engineering majors once they're in college. I'm currently working on improving 3D Printshop as it rolls into its second year, focusing on developing staff skills in 3D design, 3D scanning, and client interaction.

I'm proud to be a part of something that is having a positive impact on youth. In some ways, I'm using the opportunity that DHF gave me and paying it forward to those who come next. I will continue to benefit from these experiences long after the paychecks stop. And like any entrepreneur, I'm always looking to grow and improve my business and myself. ◐

MONEY MACHINE
Want to put your 3D printer to work? Here's how to make some cold, hard cash.

There are lots of ways to make a few bucks on the side using your 3D printer, especially if you know how to 3D model. Two tips: First, come up with something unique. People will want it more if you're the only one offering it. Second, don't forget to charge enough. Follow this equation: (materials + labor + expenses + profit) × 4 = *minimum* retail cost.

SELL YOUR FILES
Myminifactory.com, CGtrader.com, and Etsy.com let you charge others to download your STLs.

SELL PRINTS ONLINE OR IN PERSON
Sculpteo and Shapeways are 3D printing service bureaus that allow people to order a print of their models. Sell them at a local craft faire or online in an Etsy shop.

PRINT OTHER'S DESIGNS
Offer your print services on 3Dhubs.com. Always check the license. A print listed as noncommercial shouldn't be printed for money, regardless of how you frame it. Telling a lawyer that "I wasn't charging for the print — I was just charging for my time and material" doesn't cut it.

SHARE YOUR KNOWLEDGE
Run "Intro to 3D Printing" classes at your makerspace. YouTube, Patreon, and Twitch can also lead to a few dollars here and there that add up over time.

FIX 3D PRINTERS
Offer repair services on Craigslist. Folks often need assistance getting their machine running.

USE 3D PRINTERS AS A TOOL
Your 3D printer can of course serve as a tool to make something else. Try selling resin casts of your print that take 15 minutes to make instead of over 6 hours on a printer.

—Todd Blatt, custom3dstuff.com

HEAVY INDUSTRY

Written by Joseph Kowen and Terry Wohlers

Large, fast, and exotic, professional 3D printers make serious parts

JOSEPH KOWEN, an associate consultant at Wohlers Associates, holds a law degree from Hebrew University in Jerusalem, Israel, and an MBA from Case Western Reserve University.

TERRY WOHLERS is an industry consultant, analyst, author, and speaker. He is president of Wohlers Associates, an independent consulting firm he founded 30 years ago.

Patents grant inventors a monopoly until they expire. Upon expiration, everyone can apply the invention to new products without limitation. All of today's fused deposition modeling (FDM)–like desktop 3D printers can trace their technological roots to a 1992 patent by Scott Crump. Thanks to Crump's inventiveness, millions of people worldwide now have access to inexpensive versions of his original idea.

Stratasys, the company Crump founded, continues to produce industrial 3D printers based on the same concept. The top-of-the-line machine today can print parts as large as 36"×24"×36" using a range of materials. Parts made in ULTEM 9085, for example, are used extensively in aircraft from Boeing and Airbus. The list price of the system? $400,000. Yet customers continue to buy such pricey machines because of the demanding applications and material properties that their businesses require.

The average selling price (ASP) of an industrial 3D printer is $104,222, according to *Wohlers Report 2017*. The ASP for printers that produce metal parts is $566,570. Meanwhile, the ASP for desktop machines that sell for under $5,000 is $1,094. Both the low and high end share the basic process of producing parts layer by layer, but they are really two very different animals.

At the time Crump was developing the FDM process, other pioneers were inventing different methods of 3D printing.

Stereolithography, in which light is used to build parts by selectively curing photopolymer, is still a popular method among industrial users for producing a wide range of parts, large and small. With the expiration of basic patents for this process, simple vat photopolymerization printers are available today for $3,500. 3D Systems, the company to first commercialize the process, still sells large stereolithography systems for up to $990,000.

3D printers based on the powder bed fusion process use a laser to melt layers of powder to build up high-performance, functional parts. Originally developed to produce plastic pieces, the same process is now at the heart of the hottest area of additive manufacturing (AM): melting metal powder to build fully functional metal parts.

INDUSTRIAL OUTLOOK

Manufacturers see AM as a next-generation tool for production parts, a $12.8 trillion global business.

Materials: Today's 3D printers can print in a long list of plastics that includes nylons, elastomers, silicones, Kevlar, carbon fiber–filled plastics, and even biocompatible materials. Industrial companies print nickel, titanium, and precious metal parts for jet engines, dental crowns, spare parts for automobiles, and even jewelry. Ceramic parts and foundry sand for casting metal parts are also printed routinely.

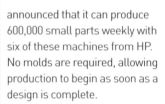

THE INDUSTRY BIBLE

Anyone with serious interest in professional 3D printing should be reading the *Wohlers Report*, an annual tome that outlines the state of AM, ranging from the largest tool makers down to the desktop machines that frequently grace the pages of *Make:*. The 2017 edition contains 343 pages of useful, up-to-date information that includes breakdowns of just about everything additive, including software options and material properties offered by printing vendors. At just under $500, it's not a casual purchase for hobbyists, but there's no better resource for anyone in the business of digital fabrication.

—*Mike Senese*

Size: The Big Area Additive Manufacturing (BAAM) system, manufactured by Cincinnati, can build parts of up to 240"×90"×72", extruding thermoplastic materials at a rate of 80 pounds per hour. One would need to run 2,725 Ultimaker 3 desktop 3D printers to reach the same volume as one BAAM system.

Throughput: Today's industrial 3D printers are capable of producing parts in quantities and at speeds that begin to compete with traditional manufacturing processes for short-run production. Wohlers Associates calculates that one would need 163 Ultimaker 3 machines to match the output of *one* HP Jet Fusion 4200 printer. A service provider in California

announced that it can produce 600,000 small parts weekly with six of these machines from HP. No molds are required, allowing production to begin as soon as a design is complete.

Applications: NASA's Marshall Space Flight Center is using metal AM to produce ignitors, injectors, combustion chambers, and turbopumps for next-generation propulsion systems. GE is using the technology to produce LEAP engine fuel nozzles that are 25% lighter, last 5 times longer, and are easier to manufacture, and the company has the capacity to manufacture tens of thousands of the 3D printed nozzles annually. GE also redesigned the engine for the CT7 helicopter, determining that it could print up to about 40% of the engine. With AM, the company was able to reduce 900 individual parts to just 16 and reduce weight and cost by around 35%.

The good news for makers is that much of what industrial system manufacturers learn today becomes public knowledge tomorrow. When patents expire, energetic entrepreneurs waste no time in packaging these processes for the broader public. In fact, it is already well underway. ●

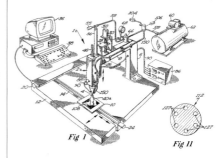

Fig 1

Fig 11

Finding the Right Fit

Danielle Applestone talks acquisition, rebranding Other Machine Co., and teaming up with **Bre Pettis**

DC DENISON is the co-editor of the *Maker Pro Newsletter*, which covers the intersection of makers and business, and is the senior editor, technology at Acquia.

Danielle Appplestone is the CEO of Bantam Tools (formerly Other Machine Co.), which makes the Bantam Tools Desktop PCB Milling Machine. During the company's first five years it raised $312,000 on Kickstarter, and $6.5 million from investors. It sold thousands of units. Recently Applestone rebranded the company and refocused its mission with new owner Bre Pettis, co-founder and former CEO of MakerBot, the pioneering consumer 3D printer manufacturer.

You've joined a small group of maker pros who have built a company and then sold it. Tell us about that.

We weren't really a good fit for a bank or more venture capital. I started having conversations with a lot of different people to see if we could find a partner. Some companies were interested in the software, but not the hardware, or vice versa, but I really wanted to find somebody who would resonate with the original vision.

When I reached out to Bre Pettis, I was actually looking for suggestions. It was a surprise that he was interested in the company.

Like many of us, you've probably read negative articles about Bre. Did that worry you?

All I knew about him was what I read in the media, so that was one of the first things I asked him about. That was really important to me: to be real with each other. I think I'm a good judge of character, and I felt at the end of the day that this person has good character. I've always believed in people's capacity for growth. And what I have experienced, in the process of him buying the company and us going though a rebranding, has further cemented my feelings about him. I'm excited to work with Bre because he has special marketing and sales skills that I don't, so in that realm he'll be hands-on. But I will still be an independent CEO.

What advice would you give to makers who find themselves in a position to sell?

If you can't get a fabulous return on your financial investment, you can get it from a "good for society" standpoint. Figure out what people want, and what will satisfy them. The other thing is: keep it moving. Over-communicate. A lot of it is just

listening, to be honest. I think that's why so many companies fail during this process. The acquisition part is hard, and it's really different from operating the business.

How do you best design a product that people want?

Build 100 prototypes, sell them, and watch people use them. You're going to learn so much in that process, like how hard it is to sell. You also feel the weight of the customer promise. You've got people using this thing, and they may grow to rely on it, and then it breaks. So you get valuable reliability testing. Probably the most important thing: you're getting real feedback from real people.

Then you can go through iterations rapid-fire. You're learning the unknowns that you're going to have to deal with later anyway.

Any advice for someone hoping to sell their product to schools?

If your product is cheap, people can just activate at the parent or teacher level. Once you get past a certain price point there will be teachers, administrators, deans, principals involved, and then there's the purchasing

department. Schools are always strapped for cash. If it's a public school, they have to really want it. In universities people have an independent budget. We sell to about 60:40 businesses:universities now.

When you rebranded, you announced that you were focusing on professional electrical engineers. Why did you pivot away from hobbyists?

Once we got past Kickstarter, it was like, "OK, who's buying?" It's a big purchase that's difficult to justify if you're a hobbyist. If you're a professional electrical engineer it's easier. Ultimately I want to grow the audience beyond that, by getting these machines into libraries. I've talked to many machine shop guys who love the Bantam Tools Milling Machine. They're saying, "I could bring it home and show my kids what I do, and how awesome it is." They understand the importance. Instead of hearing, "Oh, boring, factory," it's, "This is the future." ◗

Read the full interview at makezine.com/go/bantam.

Bantam Tools

Danielle Applestone

Bre Pettis

THE *Fine* PRINT

Our annual shootout is just the beginning of producing our detailed digital fabrication machine reviews

Written by Matt Stultz

Welcome to our 2018 Digital Fabrication Guide. This is our sixth year of testing and reviewing 3D printers and third year testing other digital fabrication tools. In that short amount of time we've seen incredible leaps in the industry, starting with wood-framed kits dominating the landscape and moving to the current pack of machines with purpose-built metal cases and injection-molded parts. It's not just the machines, either — the average 3D printer user is changing too, now expecting a higher quality, more reliable tool because they tend to be more interested in the print than the printer itself.

TESTING AND SCORING

Each year our team gathers for our shootout testing weekend, three straight days of putting these machines through our continuously improving process to try to see how they stack up against each other. For the second straight year, we held our shootout in my home hackerspace, Ocean State Maker Mill in Pawtucket, Rhode Island. The team consisted of other makerspace members who have extensive experience with 3D printing, CNC milling, laser cutting, and other digital fabrication processes.

For the fused filament fabrication (FFF) machines, the initial testing is just the start, leading to blind print scoring of our test probes, comparing our notes, and really diving into the details of the machines to find the standouts.

Our Digital Fabrication Guide is one of most popular issues every year. We continue to try to improve our tests based on reader feedback and are always looking for new machine types that we think you will be interested in. I hope you enjoy this year's guide and that it will help you find that special tool you've been looking for. ⊘

MEET THE TESTERS

PHILIP J. ANGILERI has been an industrial designer for over 20 years. He is president and principal for design at NarrowBase LLC and is a member of Ocean State Maker Mill.

ADAM CASTO is a Linux sysadmin and maker. He is an active member of HackPGH and 3DPPGH in Pittsburgh, PA.

MATT DAURAY is a mechanical engineer and materials connoisseur. A finish carpenter and leatherworker, Matt spends his spare time at the Ocean State Maker Mill.

KELLY EGAN is an artist, teacher, and creative coder based in Providence, R.I. He is a founding member of the Baltimore Node Hackerspace and Ocean State Maker Mill.

CHAD ELISH is a driving force in the maker movement. He's the president of HackPGH, produces Maker Faires throughout the U.S., and co-founded Nation of Makers.

DARIUS MCCOY is the 3D printing manager at the Digital Harbor Foundation where he founded 3D Assistance and the new Print Shop, a youth-run 3D printing service.

SIMON NORRIDGE has been a maker since building his first computer in the early 1970s. He now focuses on creating with CNC and 3D printing systems at Ocean State Maker Mill.

RYAN PRIORE is a spectroscopist and entrepreneur in the photonics industry. He is also the chief bed leveler of 3DPPGH and an active member of HackPGH.

JONATHAN PROZZI is the director of education at the Digital Harbor Foundation where he develops technology content and resources for youth and educators.

JEN SCHACHTER is a fellow at RWD Foundation studying maker culture, spaces, and education. She is a frequent collaborator with tested.com, and resident artist at Open Works Baltimore.

MANDY L. STULTZ works in online marketing and helps run Ocean State Maker Mill. She is a collector of shoes and vintage two-wheeled vehicles, and a keeper of four-footed beasts.

MATT STULTZ is *Make:*'s digital fabrication editor in charge of heading up this team and is the founder of 3DPPVD, Ocean State Maker Mill, and HackPGH.

CHRIS YOHE is co-founder of 3DPPGH and a HackPGH member. Software developer. Hardware hacker. Rugby player. He has an undisclosed number of manufacturing minions.

Hep Svadja, Jukka Seppänen

TEST PRINTS:
Explained

Illustrated by Rob Nance

Our FFF test probes allow us to quantify how well the printers will produce parts. Here is a breakdown of the tests we use and how we score them.

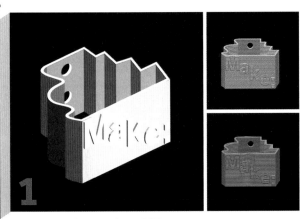

VERTICAL SURFACE TEST

With this probe, we look for echoing or banding on the surface of the print. These will be located along the letters or the holes in the back. The worse the disruption, the lower the score.

OVERHANG TEST

The overhang test starts out at 30º and continues to 70º. We look for how cleanly the back of the print comes out. Drooping and looping especially in the lower numbers will reduce the score.

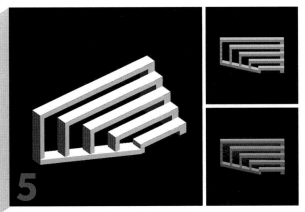

BRIDGING TEST

Bridging is a printer's ability to span a gap between two points. Each gap on this probe is wider apart than the last. A good part should have all the gaps bridged cleanly. Drooping and lost infill reduces points.

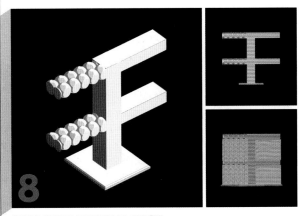

SUPPORT MATERIAL TEST

This is an updated test for this year. This tree tests how well support material is formed to a print. It tests four conditions that are a combination of flat and textured print areas. The print is scored based on how cleanly the material can be removed.

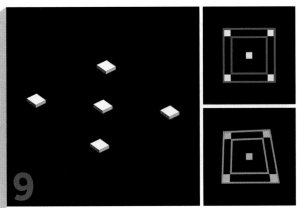

SQUARENESS TEST

This test is one of two new tests for 2018. Five squares are printed around the bed, each are measured with an angle gauge to see how close they are to 90º. The further off they are from 90º, the lower the score.

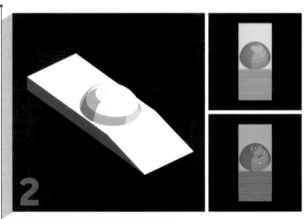

2 HORIZONTAL FINISH TEST

There are three sections to this test, a slope, a flat section, and a dome. We are looking to see how cleanly those sections are formed. Any holes, pimples, or ridging will lower the score.

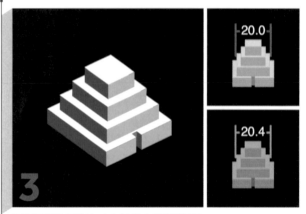

3 DIMENSIONAL ACCURACY TEST

This probe allows us to measure how accurately the printer produces parts to size. The second step should be 20mm wide in both directions. For each .1mm the printer is off on average, it loses a point.

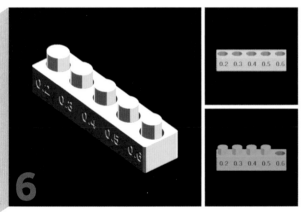

6 NEGATIVE SPACE TOLERANCE TEST

Negative space tolerance is important for things like bolt holes and creating print-in-place features. Each hole in the test is .1mm smaller than the last. The more pins that can be pushed out, the higher the score.

7 RETRACTION PERFORMANCE TEST

Poor retraction can cause stringing, jamming, or the inability to extrude fine features. These spikes test your machines retraction performance. Any of the issues listed can reduce the score of the print.

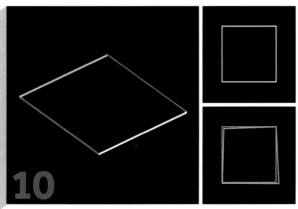

10 FULL BED DIMENSIONAL ACCURACY TEST

This is our other new test this year. We heard feedback from our readers who wanted to make sure the printers could print out to the edges and also stayed accurate with big prints. We measure this print and the less accurate, the lower the score.

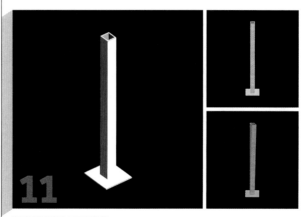

11 Z WOBBLE TEST

This tower tests to see if there is any wiggle in your z-axis causing repeated ridging along the sides. This is our lone pass/fail test now and scores a 0 (fail) or 2 (pass).

RAISE3D N2
Written by Philip J. Angileri

Great features plus integrated software and hardware put this machine at the front of the pack

ANYBODY KEEPING SCORE OVER THE PAST FEW YEARS CAN SEE THAT IT'S GETTING LONELY AT THE $3K MARK. But you are certainly getting what you pay for when it comes to the N2. The display coupled with the slick proprietary software, dual extrusion, and massive build volume keep this machine atop the competition.

ATTRACTIVE AND ENCLOSED

Quiet enough for the office, handsome enough for the conference room. The full enclosure will protect it in a dusty environment as well as cut down on the variables that things like AC units or open windows can introduce. The touchscreen interface was the best one at the event for me: The display is large and the menus were easy to navigate. The N2 is definitely not for someone into open source alternatives, hacking, or customizing, but that does allow you to spend less time tuning and more time designing.

The bed could be improved — it's a pain to level. The company website states that it comes pre-calibrated from the factory, however ours was not. The build area is large, but watch out for the binder clips holding the bed in place, something I think could be better handled with another solution, especially at this price range.

WORKS LIKE A CHARM

There is something to be said about integrated software and hardware. The N2 software is proprietary, but that is a major factor in why it is so good out of the box. It really does just work. I had no connection issues, no software issues. Get your bed flat, be careful where you place that binder clip. That's it.

LEADER OF THE PACK

Even a week after the shootout I was still blown away by how far the display and ideaMaker slicing software set this machine apart from the rest. Custom slicers appear to be a trend now and Raise3D has been doing it for a while, making them a safe bet. If you've been considering buying the N2, you might want to spend the extra money and get the N2 Plus instead for the foot extra build height. You deserve it. ◉

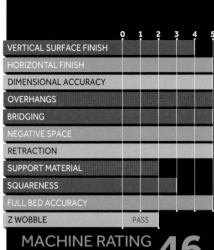

	0	1	2	3	4	5
VERTICAL SURFACE FINISH						
HORIZONTAL FINISH						
DIMENSIONAL ACCURACY						
OVERHANGS						
BRIDGING						
NEGATIVE SPACE						
RETRACTION						
SUPPORT MATERIAL						
SQUARENESS						
FULL BED ACCURACY						
Z WOBBLE			PASS			

MACHINE RATING 46
Price as Tested **$2,999**

- **WEBSITE** raise3d.com
- **MANUFACTURER** Raise3D
- **BUILD VOLUME** 305×305×305mm (single and dual extrusion)
- **BED STYLE** Heated glass with BuildTak
- **FILAMENT SIZE** 1.75mm
- **OPEN FILAMENT?** Yes
- **TEMPERATURE CONTROL?** Yes, extruder (300°C max); bed (110°C max)

- **PRINT UNTETHERED?** Yes, Wi-Fi, LAN, USB, and SD card. Has print resume for power outage protection.
- **ONBOARD CONTROLS?** Yes, 7" touchscreen
- **HOST/SLICER SOFTWARE** ideaMaker custom slicer
- **OS** Mac, Windows, Linux
- **FIRMWARE** Vendor provided
- **OPEN SOFTWARE?** No
- **OPEN HARDWARE?** No

PRO TIPS
Despite lots of fine tuning, our test bed never really got super flat. I ran most of my test prints with rafts.

Guys. Binder clips are not hardware. Let's all stop pretending it's acceptable on a production machine.

WHY TO BUY
This thing is built for the living room or a man cave. It's quiet, completely enclosed in plexiglass, and almost the perfect height for a side table. If you decide to go that route, use a coaster.

Make: BEST OVERALL N2 Raise3D

Make: BEST LARGE FORMAT N2 Raise3D

TEST PRINT

Hep Svadja

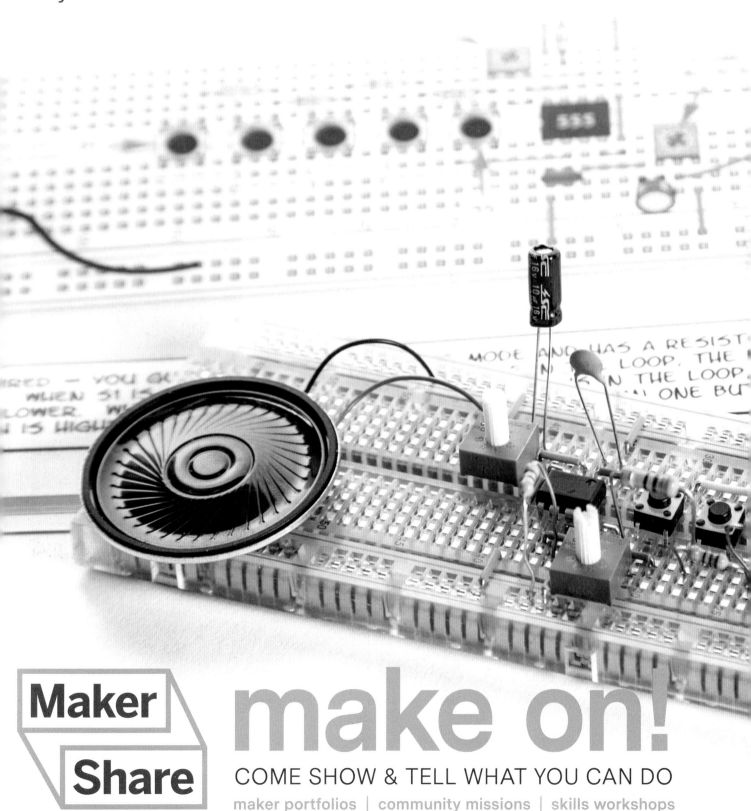

i3 MK2S

Advancements make last year's top-rated machine even better Written by Chad Elish

LAST YEAR THE PRUSA I3 MK2 TOPPED THE LIST AS OUR BEST MACHINE. This year we put the upgraded MK2S to the test, and it shared the No. 1 ranking.

UPGRADED HARDWARE

One of the highlights of this machine is the bed and how it handles leveling. Before every print the MK2S runs a quick measurement of 9 points on the plate with the PINDA probe and adjusts the print in real time to compensate for any out of level or out of square measurements. The live z adjustment was our favorite feature of the calibration procedure — with a click and spin of the knob you can easily change the z height, and it actually remembers the setting.

One major improvement is the removal of the zip ties holding the bed in place. Metal U bolts hold the bearings in place, which greatly helps the overall rigidity of the machine and ensures that every time you go to print it's ready for you. Already own a Prusa i3 plus or MK2? Prusa Research offers upgrade options at a reasonable price.

UPDATED SOFTWARE

The new slicer and machine control software, PrusaControl, is designed for someone who just wants to click and print. Those of you without dual extruders will enjoy the built-in filament color change feature: Simply use a slider to choose at what layer you want to swap and with a click of the mouse, a pause will be added to the print allowing you to easily change the filament manually. Want to dive in a bit deeper? Like its predecessors, the MK2S is open and gives you freedom.

PROVEN PERFORMANCE

With a genuine E3D V6 all-metal hot end you can print with virtually any filament — it can reach temperatures of up to 300°C. The E3D V6 is considered by many to be the best hot end on the market, and its inclusion in the Prusa MK2 line of printers is one of the reasons they rise to the top. And with the MK3 hitting the market, the MK2S' price has dropped $100, starting at $599 for the kit. ⊘

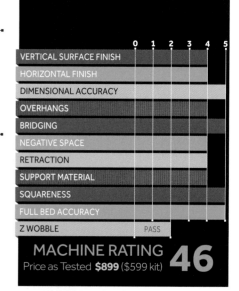

	0	1	2	3	4	5
VERTICAL SURFACE FINISH						
HORIZONTAL FINISH						
DIMENSIONAL ACCURACY						
OVERHANGS						
BRIDGING						
NEGATIVE SPACE						
RETRACTION						
SUPPORT MATERIAL						
SQUARENESS						
FULL BED ACCURACY						
Z WOBBLE	PASS					

MACHINE RATING 46
Price as Tested **$899** ($599 kit)

- **WEBSITE**
  prusa3d.com
- **MANUFACTURER**
  Prusa Research
- **BUILD VOLUME**
  250×210×200mm
- **BED STYLE** Heated with PEI print surface
- **FILAMENT SIZE**
  1.75mm
- **OPEN FILAMENT?**
  Yes
- **TEMPERATURE CONTROL?** Yes, extruder (300°C max); bed (120°C max)
- **PRINT UNTETHERED?**
  Yes, SD card

- **ONBOARD CONTROLS?**
  Yes, clickable scroll wheel with LCD
- **HOST/SLICER SOFTWARE**
  PrusaControl and custom Prusa Slic3r
- **OS** Mac, Windows, Linux
- **FIRMWARE**
  Marlin
- **OPEN SOFTWARE?**
  Yes, GNU GPLv3
- **OPEN HARDWARE?**
  Yes, GNU GPLv3

PRO TIPS
If you're having issues with prints not sticking to the bed, wipe it off with alcohol, run a z calibration, and use the live z adjustment.

WHY TO BUY
The Prusa i3 MK2 was our top rated printer last year, and the MK2S upgrades made the best better. A price drop when the MK3 was released just sweetens the deal.

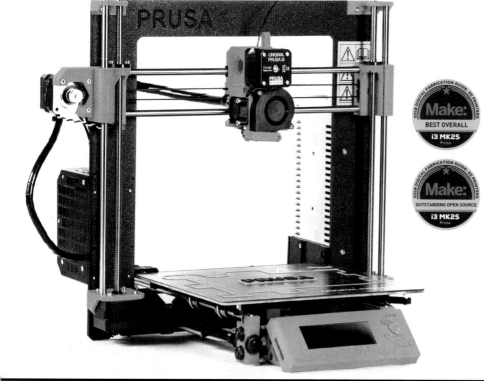

Hep Svadja

TEST PRINT

i3 MK2/S MULTI MATERIAL
Written by Ryan Priore

Step up to 4-color printing with this easy, inexpensive kit

PRUSA'S MULTI MATERIAL UPGRADE FOR THE I3 MK2/S PRINTER brings 4-color printing without compromising the single color printing quality and workflow.

SMOOTH TRANSITIONS

You don't have to commit to a particular extruder when slicing, which makes switching between colors as simple as selecting the appropriate extruder number in software. But be prepared: Multimaterial printing is a long process, especially by 3D printing terms.

Two nitpicky comments: the amount of wasted filament as part of the purge plate process, and upon canceling a print the filament is expelled by default.

GREAT VALUE, BROAD APPEAL

The MK2S MM printer blew through our testing probes and then went on to dazzle us with its multicolor tricks. At a mere $299, the MM upgrade is a no-brainer. ⊘

WHY TO BUY
The Multi Material upgrade is the most economical method of bringing 4-color 3D printing to an already powerhouse machine.

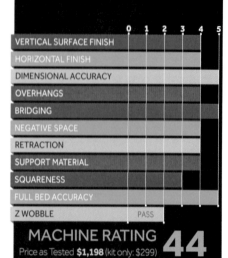

	0	1	2	3	4	5
VERTICAL SURFACE FINISH						
HORIZONTAL FINISH						
DIMENSIONAL ACCURACY						
OVERHANGS						
BRIDGING						
NEGATIVE SPACE						
RETRACTION						
SUPPORT MATERIAL						
SQUARENESS						
FULL BED ACCURACY						
Z WOBBLE			PASS			

MACHINE RATING 44
Price as Tested **$1,198** (kit only: $299)

- **WEBSITE** prusa3D.com
- **MANUFACTURER** Prusa Research
- **BUILD VOLUME** 250×210×200mm (quad extrusion)
- **BED STYLE** Heated with PEI print surface
- **FILAMENT SIZE** 1.75mm
- **OPEN FILAMENT?** Yes
- **TEMPERATURE CONTROL?** Yes, extruder (300°C max); bed (120°C max)
- **PRINT UNTETHERED?** Yes, SD card
- **ONBOARD CONTROLS?** Yes, clickable scroll wheel with LCD
- **HOST/SLICER SOFTWARE** PrusaControl and custom Prusa Slic3r
- **OS** Mac, Windows, Linux
- **FIRMWARE** Marlin
- **OPEN SOFTWARE?** Yes, GNU GPLv3
- **OPEN HARDWARE?** Yes, GNU GPLv3

---

i3 MK3
Written by Matt Stultz

Prusa's newest machine is poised to outshine its siblings

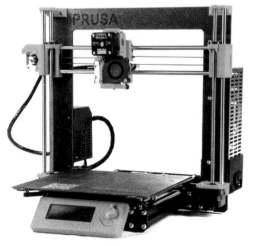

THE PRUSA TEAM COULD'VE BEEN SATISFIED WITH LAST YEAR'S TOP SPOT — instead, they pushed to keep improving.

LOTS TO LOVE

The MK3 finally trades threaded rods for 30mm aluminum extrusion, and strengthened cable mounts help prevent wear and tear. The upgraded RAMBo Einsy controller sports new stepper drivers: Trinamic TMC2130s, which enable it to detect a stalled motor or missing steps, eliminating the need for endstops. Extruder monitoring lets it pause for jams and gaps, and the bed is now a magnetically attached spring steel sheet bonded with PEI.

PURSUING EXCELLENCE

Our test prints were incredible, but the MK3 fell behind its sibling. However, our untuned prototypes were the only two in existence at the time of testing, so the score likely will rise with future testing of the final version. ⊘

WHY TO BUY
The new MK3's spec list shows the Prusa team is not just trying to please their customers, they want a great machine for themselves as well — and will continue to make it better.

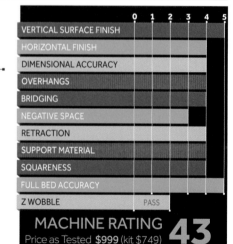

	0	1	2	3	4	5
VERTICAL SURFACE FINISH						
HORIZONTAL FINISH						
DIMENSIONAL ACCURACY						
OVERHANGS						
BRIDGING						
NEGATIVE SPACE						
RETRACTION						
SUPPORT MATERIAL						
SQUARENESS						
FULL BED ACCURACY						
Z WOBBLE			PASS			

MACHINE RATING 43
Price as Tested **$999** (kit $749)

- **WEBSITE** prusa3d.com
- **MANUFACTURER** Prusa Research
- **BUILD VOLUME** 250×210×200mm
- **BED STYLE** Heated, removable, PEI coating
- **FILAMENT SIZE** 1.75mm
- **OPEN FILAMENT?** Yes
- **TEMPERATURE CONTROL?** Yes, extruder (300°C max); bed (120°C max)
- **PRINT UNTETHERED?** Yes, SD card and Wi-Fi via custom integrated OctoPrint
- **ONBOARD CONTROLS?** Yes, clickable scroll wheel with LCD
- **HOST/SLICER SOFTWARE** PrusaControl and custom Prusa Slic3r
- **OS** Mac, Windows, Linux
- **FIRMWARE** Marlin
- **OPEN SOFTWARE?** Yes, GNU GPLv3
- **OPEN HARDWARE?** Yes, GNU GPLv3

Hep Svadja

PRINTRBOT SIMPLE PRO

A trusted fave attempts a new streamlined workflow

Written by Chris Yohe

WITH THE PRINTRBOT SIMPLE PRO, BROOK DRUMM AND TEAM are trying to deliver what he believes is his easiest to use machine yet. While we didn't have a totally smooth experience with it, you will find the machine quality that one expects from Printrbot, and a whole new way of printing to go alongside the old standards.

NEW AND IMPROVED

The Simple Pro combines a color touchscreen with the new Printrboard G2, a 32-bit controller running G2 Core (the ARM port/successor to TinyG). These tie it into Printrbot's new cloud software and ideally allow one of the easiest printing experiences yet: a true scroll, click, and print process that lets you, with as few settings as possible, send your STLs to the machine and have them print. This means no more separate slicing software hassles.

WOES WITH WOWS

Through the shootout weekend the team had issues getting Printrbot's cloud software to work smoothly, so we used Cura, tethered via USB, to generate our test prints. Assessing the quality of these probes, the testing team found that the machine was capable of producing great results. Once I got the Wi-Fi and cloud set up and running I had small hiccups, like machine freezes during preheating and buttons missing on the LCD. These were usually remedied with a good old fashioned reboot. But overall my experience went fine, and to be able to load designs as I found them, from anywhere, and print them when I get home was refreshing, and certainly kept my downloads folder cleaner.

HAS POTENTIAL

It would be nice to have some of the edges ironed out sooner rather than later, especially the need for regular restarts. While these rough spots are irritating, this is a machine that could appeal to users who don't want to get bogged down in settings and slicers. As the software catches up with the potential of the hardware, we will see smoother experiences in the future. ◐

	0	1	2	3	4	5
VERTICAL SURFACE FINISH						
HORIZONTAL FINISH						
DIMENSIONAL ACCURACY						
OVERHANGS						
BRIDGING						
NEGATIVE SPACE						
RETRACTION						
SUPPORT MATERIAL						
SQUARENESS						
FULL BED ACCURACY						
Z WOBBLE	PASS					

MACHINE RATING 43
Price as Tested **$699**

- **WEBSITE**
  printrbot.com
- **MANUFACTURER**
  Printrbot
- **BUILD VOLUME**
  200×150×200mm
- **BED STYLE**
  Heated, with removable plate
- **FILAMENT SIZE**
  1.75mm
- **OPEN FILAMENT?**
  Yes
- **TEMPERATURE CONTROL?**
  Yes, extruder (270°C max), bed (100°C max)
- **PRINT UNTETHERED?**
  Yes, Wi-Fi via Printrbot.cloud

- **ONBOARD CONTROLS?**
  Yes, full color LCD touchscreen
- **HOST/SLICER SOFTWARE**
  Printrbot.cloud, Cura
- **OS** Mac, Windows, Linux
- **FIRMWARE**
  Custom G2 Core
- **OPEN SOFTWARE?**
  Yes, Cura is AGPLv3. Printrbot has also open sourced their cloud and hub with MIT License
- **OPEN HARDWARE?**
  Yes, CC-BY-SA.

PRO TIPS
Issues like machine freezes, the cloud stuttering, and buttons missing on the LCD were easily remedied with a good old-fashioned reboot.

WHY TO BUY
Rigid linear rails for the y-axis and a magnetic build plate are standard.

The expected Printrbot print quality is now packaged with a full wireless print solution and touchscreen for an all-in-one package.

TEST PRINT

Hep Svadja

printrbot

FELIX 3.1 Written by Philip J. Angileri

This sturdy, accurate machine is great for schools or professionals

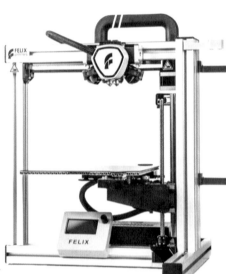

I'VE SEEN A LOT OF HIGH-PERFORMING, SUB-$1,000 PRINTERS ON THE MARKET, but what sets apart $2,000 machines — like the Felix 3.1 — is robust construction.

SOLID STRUCTURE
Linear rails add stability. I wouldn't think twice about running this machine at higher speeds. The high-temp, direct drive extruders are also suitable for materials beyond PLA and ABS.

SOLID PRINTING
Felix 3.1 produced good results with no jams or skips, and the option for dual extrusion opens up the opportunity for separate support material. Plus, it was quiet for an open-air design. While the price might put off some hobbyists, it's clear Felix is targeting professionals and schools with the construction, upgradeability, lifetime support, and easy portability. ⊘

WHY TO BUY
If your budget is in the $2K range, the Felix 3.1's bulletproof construction, reliability, and customer service make it an attractive buy.

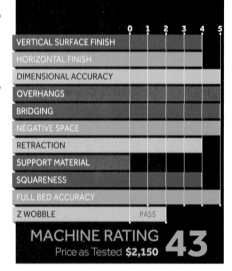

	0	1	2	3	4	5
VERTICAL SURFACE FINISH						
HORIZONTAL FINISH						
DIMENSIONAL ACCURACY						
OVERHANGS						
BRIDGING						
NEGATIVE SPACE						
RETRACTION						
SUPPORT MATERIAL						
SQUARENESS						
FULL BED ACCURACY						
Z WOBBLE	PASS					

MACHINE RATING 43
Price as Tested **$2,150**

- **WEBSITE** felixprinters.com
- **MANUFACTURER** Felix Printers
- **BUILD VOLUME** 255×205×225mm (single extrusion) 240×205×225mm (dual extrusion)
- **BED STYLE** Heated with Kapton tape
- **FILAMENT SIZE** 1.75mm
- **OPEN FILAMENT?** Yes
- **TEMPERATURE CONTROL?** Yes, extruder (275°C max); bed (95°C max)
- **PRINT UNTETHERED?** Yes, USB, SD card
- **ONBOARD CONTROLS?** Yes, single control knob (rotary/push)
- **HOST/SLICER SOFTWARE** Simplify 3D (recommended; vendor provided custom profile), Felixbuilder (yearly subscription), Repetier-Host
- **OS** Mac, Windows, Linux
- **FIRMWARE** Vendor provided
- **OPEN SOFTWARE?** No
- **OPEN HARDWARE?** No

HACKER H2 Written by Ryan Priore

A delta printer for the consummate tinkerer

SEEMECNC'S HACKER H2 KIT ISN'T FOR THE FAINT OF HEART, but it will satisfy users who like getting their hands dirty customizing their printer.

OPEN ARMS
The kit ships with two different length sets of delta arms: Orion (178mm) and Rostock (290mm). By swapping the arms, you can get either taller or wider prints. A simple G-code macro will update the necessary EEPROM settings for the set of arms in use.

PRINT YOUR WAY
The H2 knocked out our test probes with ease. However, too little cooling at the nozzle exit resulted in poor print performance on features like overhangs. Print quality was noticeably superior with Cura over MatterControl.

Though novice users may want to look elsewhere, the H2 offers a 3D printer modder's dream at a reasonable price. ⊘

WHY TO BUY
Those who enjoy customizing their printing rig will love the build volume reconfigurability, quality components, onboard controls, auto bed leveling, and reasonable price tag.

	0	1	2	3	4	5
VERTICAL SURFACE FINISH						
HORIZONTAL FINISH						
DIMENSIONAL ACCURACY						
OVERHANGS						
BRIDGING						
NEGATIVE SPACE						
RETRACTION						
SUPPORT MATERIAL						
SQUARENESS						
FULL BED ACCURACY						
Z WOBBLE	PASS					

MACHINE RATING 42
Price as Tested **$549**

- **WEBSITE** seemecnc.com
- **MANUFACTURER** SeeMeCNC
- **BUILD VOLUME** 175mm (dia.)×200mm (tall) OR 140mm (dia.)×295mm (tall)
- **BED STYLE** Non-heated glass
- **FILAMENT SIZE** 1.75mm
- **OPEN FILAMENT?** Yes
- **TEMPERATURE CONTROL?** Yes, extruder (280°C max)
- **PRINT UNTETHERED?** Yes, SD card
- **ONBOARD CONTROLS?** Yes, scroll knob and LCD
- **HOST/SLICER SOFTWARE** MatterControl (vendor recommended), Cura
- **OS** Mac, Windows, Linux
- **FIRMWARE** Repetier-Firmware
- **OPEN SOFTWARE?** Yes, MatterControl is BSD-2-Clause, Cura is AGPLv3
- **OPEN HARDWARE?** Yes, GNU GPL 3.0

Hep Svadja

ULTIMAKER 3 — Written by Jonathan Prozzi

The company continues to improve on their already high-quality printer with some new additions

THE ULTIMAKER 3 RETAINS THE PRECISE, RELIABLE PRINTS of its predecessors, while adding dual extrusion, Wi-Fi/LAN, and active bed leveling. Since Ultimaker is firmly established, there's a wealth of information and support available.

WHAT'S NEW

The dual extruders are a welcome addition. One application is to use the AA nozzle to extrude PLA for the main body and then the BB nozzle for water-soluble PVA support. No need to chisel away PLA support material.

The swappable print cores are another notable feature. Changing nozzles is beginner friendly, and the LCD guides you through it. In addition to printing with multiple materials, educators could keep a spare print core around to swap in when the nozzle clogs (one of the most common issues across all brands). You can purchase a .8mm nozzle for faster print time, as Ultimaker's default settings are tailored for quality over speed.

The machine now auto-detects (via NFC) Ultimaker filament, setting the material and color on the machine and in Cura.

The Ultimaker 3 can print untethered via USB or Wi-Fi. The Wi-Fi configuration is as seamless as everything else, and the user can send and monitor prints with its new built-in camera via Cura without needing to hop off their regular wireless network.

SOME CONFUSION

While Ultimaker's specs state that the build volume is 215×215×200mm for single extrusion and 197×215×200mm for dual extrusion, the *actual* build volume allowable with Cura's defaults is 194×182mm (single) and 176×182mm (dual) because of bed-clip dodge areas.

A GREAT BUY OVERALL

The Ultimaker 3 is a powerful, reliable, high-quality, and versatile machine with a wide audience. While the price is somewhat high, you're receiving a reputable product. I'd recommend this for nearly everyone with the ability to afford it. ◗

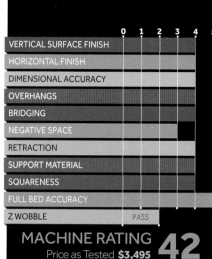

	0	1	2	3	4	5
VERTICAL SURFACE FINISH						
HORIZONTAL FINISH						
DIMENSIONAL ACCURACY						
OVERHANGS						
BRIDGING						
NEGATIVE SPACE						
RETRACTION						
SUPPORT MATERIAL						
SQUARENESS						
FULL BED ACCURACY						
Z WOBBLE	PASS					

MACHINE RATING 42
Price as Tested **$3,495**

- **WEBSITE**
  ultimaker.com
- **MANUFACTURER**
  Ultimaker
- **BUILD VOLUME**
  194×182×200mm (single extrusion)
  176×182×200mm (dual extrusion)
- **BED STYLE**
  Heated glass
- **FILAMENT SIZE**
  2.85mm
- **OPEN FILAMENT?**
  Yes
- **TEMPERATURE CONTROL?**
  Yes, extruder (280°C max); bed (100°C max)
- **PRINT UNTETHERED?**
  Yes, Wi-Fi, USB
- **ONBOARD CONTROLS?**
  Yes, LCD screen with analog wheel for selections
- **HOST/SLICER SOFTWARE**
  Cura Ultimaker edition
- **OS** Mac, Windows, Linux
- **FIRMWARE**
  Available via onboard update
- **OPEN SOFTWARE?**
  Yes, Cura is AGPLv3
- **OPEN HARDWARE?**
  Yes, CC-BY-NC

PRO TIPS
Take advantage of how quick and easy it is to change print cores. You're able to swap in spare nozzles to address clogs, print with multiple material types, and increase the speed by using a larger diameter hot end.

WHY TO BUY
The Ultimaker 3 retains the high quality of past iterations while adding versatility with dual extrusion, Wi-Fi, and swappable print cores. It supports a variety of materials with optimized Cura profiles and settings.

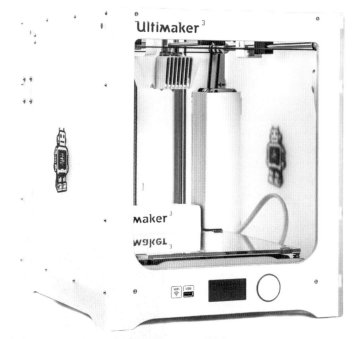

Hep Svadja

TEST PRINT

FELIX PRO 2

An approachable dual extrusion machine   Written by Kelly Egan

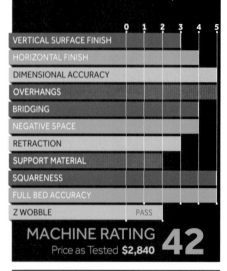

	0	1	2	3	4	5
VERTICAL SURFACE FINISH						
HORIZONTAL FINISH						
DIMENSIONAL ACCURACY						
OVERHANGS						
BRIDGING						
NEGATIVE SPACE						
RETRACTION						
SUPPORT MATERIAL						
SQUARENESS						
FULL BED ACCURACY						
Z WOBBLE		PASS				

MACHINE RATING   42
Price as Tested **$2,840**

WHILE THE FELIX PRO 2 IS MORE EXPENSIVE THAN MANY, IT COMES WITH lots of features to make printing faster, more accurate, and more pleasant.

MIND THE GAPS
The easy to run, four-point auto-leveling system makes sure the new bed is calibrated and ready to go. Prints were generally good but issues with temporary extruder jams left small gaps. The flow detection system occasionally warned us of jams, but none were long enough for the machine to automatically pause the print. The Pro 2 also sports a second extruder for dual printing, which tucks out of the way when not in use.

EASE OF USE
Overall the Felix Pro is a well-put-together machine with great new functions to make the printing process smoother, particularly for beginners. ⊘

WHY TO BUY
Great new features make the printing process smooth, especially for beginners or users who just need their printers to work.

- **WEBSITE** felixprinters.com
- **MANUFACTURER** Felix Printers
- **BUILD VOLUME** 237×244×235mm (single and dual extrusion)
- **BED STYLE** Heated, Kapton tape, removable
- **FILAMENT SIZE** 1.75mm
- **OPEN FILAMENT?** Yes
- **TEMPERATURE CONTROL?** Yes, extruder (275°C max); bed (100°C max)
- **PRINT UNTETHERED?** Yes, SD card
- **ONBOARD CONTROLS?** Yes, LCD display
- **HOST/SLICER SOFTWARE** Simplify3D recommended, also Felixbuilder or Repetier-Host
- **OS** Mac, Windows, Linux
- **FIRMWARE** Repetier-Firmware
- **OPEN SOFTWARE?** No
- **OPEN HARDWARE?** No

CRAFTBOT XL

A reliable, easy-to-use printer   Written by Darius McCoy

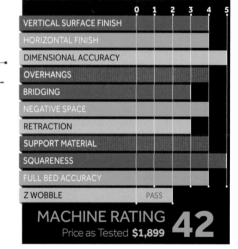

	0	1	2	3	4	5
VERTICAL SURFACE FINISH						
HORIZONTAL FINISH						
DIMENSIONAL ACCURACY						
OVERHANGS						
BRIDGING						
NEGATIVE SPACE						
RETRACTION						
SUPPORT MATERIAL						
SQUARENESS						
FULL BED ACCURACY						
Z WOBBLE		PASS				

MACHINE RATING   42
Price as Tested **$1,899**

THE CRAFTBOT XL IS A GOOD PLUG-AND-PLAY SOLUTION. The printing quality was great and the prints were fairly quick.

MATERIAL OPTIONS
The XL comes equipped with a .04 nozzle with a max resolution of 100 microns. The XL uses the company's CraftWare slicing software, and slicing files was easy and straightforward. The XL can handle common materials like ABS and PLA, plus advanced materials like HIPS, Nylon, and PETG. I would definitely want to see some improvements such as adding an auto-level feature and increasing the build plate size.

STRESS-FREE OPERATION
At $1,899, the XL's price may be a bit steep for some users, but it's a good purchase for someone looking to print without worrying about failures.

WHY TO BUY
It's a big, capable machine that produces quality prints, worry-free.

- **WEBSITE** craftunique.com
- **MANUFACTURER** Craft Unique
- **BUILD VOLUME** 300×200×440mm
- **BED STYLE** Heated with PEI surface
- **FILAMENT SIZE** 1.75mm
- **OPEN FILAMENT?** Yes
- **TEMPERATURE CONTROL?** Yes, extruder (260°C max); bed (110°C max)
- **PRINT UNTETHERED?** Yes, Wi-Fi, USB
- **ONBOARD CONTROLS?** Yes, color LCD touchscreen
- **HOST/SLICER SOFTWARE** CraftWare
- **OS** Mac, Windows, Linux
- **FIRMWARE** Proprietary
- **OPEN SOFTWARE?** No
- **OPEN HARDWARE?** No

MAKERGEAR M3

This rugged and reliable machine is as close to "it just prints" as it gets Written by Chris Yohe

THE LATEST MAKERGEAR OFFERING IS ITS NEW FLAGSHIP PRODUCT, THE M3. The steel frame combined with CNC-machined aluminum parts riding on linear rails provides a locked-in motion system that is rugged and reliable. New features include an upgraded hot end, the possibility of independent dual extrusion, and a built-in custom OctoPrint installation that comes stock on all models.

DURABLE AND DEPENDABLE

The new hot end design worked well and was able to reach 300°C, printing various filament types without any further modifications. New users will find a helpful setup script that walks you step-by-step through a great bed-leveling tutorial using the included web interface. The heated bed shined with its easily removable and swappable glass and PEI surface. The machine steamed through our probes with good results, and the overnight print was one of the best of the bunch. We do recommend removing the glass bed before removing your parts to allow faster cooling, as the bed can work a little too well at holding prints.

OCTOPRINT ONBOARD

The real star of this machine is the custom OctoPrint install. With the M3, MakerGear has delivered an easy, out of the box print experience that worked great for sending and slicing files. Advanced users can still modify their settings, but the stock settings netted great results. While we had a few Wi-Fi configuration blips, this was one of the smoother Pi-based printing experiences we've had, and one of the first built-in ones we can recommend. The ability to use it from your mobile device makes it even better.

We expect to see the web interface updated to make it even more seamless, and a little extra fail-safe for a bad Wi-Fi setup will make a good printer even better. Also a few tweaks to the default minimum layer settings will lead to a much smoother surface finish.

AN IMPRESSIVE MACHINE

Whether you are looking for a printer to crank out engineering prototypes or highly detailed dragon models, the MakerGear M3 will deliver a quality experience to your desktop that you can literally lean on. ◐

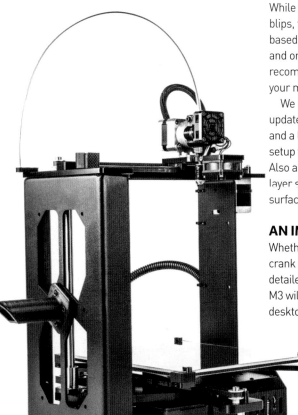

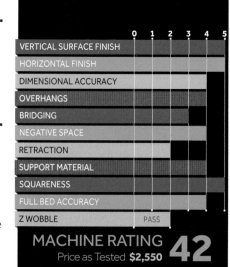

	0	1	2	3	4	5
VERTICAL SURFACE FINISH						
HORIZONTAL FINISH						
DIMENSIONAL ACCURACY						
OVERHANGS						
BRIDGING						
NEGATIVE SPACE						
RETRACTION						
SUPPORT MATERIAL						
SQUARENESS						
FULL BED ACCURACY						
Z WOBBLE			PASS			

MACHINE RATING 42
Price as Tested **$2,550**

WEBSITE
makergear.com

MANUFACTURER
MakerGear

BUILD VOLUME
203×254×203mm

BED STYLE
Heated, removable glass surface with PEI

FILAMENT SIZE
1.75mm

OPEN FILAMENT?
Yes

TEMPERATURE CONTROL?
Yes, extruder (300°C) ; bed (130°C)

PRINT UNTETHERED?
Yes, comes with OctoPrint built in from manufacturer, Wi-Fi, LAN, USB

ONBOARD CONTROLS?
No

HOST/SLICER SOFTWARE
OctoPrint onboard

OS
Mac, Windows, Linux

FIRMWARE
Modified Marlin

OPEN SOFTWARE?
Yes, GNU GPL v3.0

OPEN HARDWARE?
Yes, CC-BY-SA 3.0

PRO TIPS

A 60-second-plus hold on the reset button will get it back to stock — for when you need to undo your latest tweaks.

The bed heats exceptionally well! An extra plate will keep you quickly printing back to back.

WHY TO BUY

MakerGear has released a solid, rugged, dialed-in printing machine that cranks out high-quality prints with a self-contained cloud that gets closer to the "it just prints" experience than we've seen.

TEST PRINT

Hep Svadja

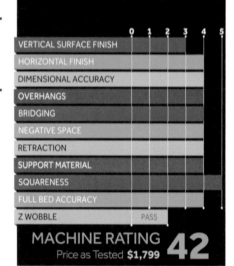

DREMEL 3D45

An IoT printer robust enough for the classroom   Written by Matt Dauray

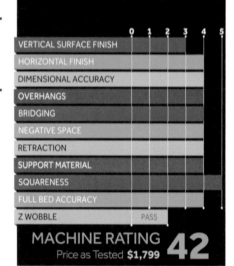

makezine.com/go/fab-guide-2018

	0	1	2	3	4	5
VERTICAL SURFACE FINISH						
HORIZONTAL FINISH						
DIMENSIONAL ACCURACY						
OVERHANGS						
BRIDGING						
NEGATIVE SPACE						
RETRACTION						
SUPPORT MATERIAL						
SQUARENESS						
FULL BED ACCURACY						
Z WOBBLE		PASS				

MACHINE RATING 42
Price as Tested **$1,799**

THE DREMEL 3D45 IS A NEW AND IMPROVED VERSION of last year's 3D40. It has the same build space and physical size with more features and functionality.

FRESH FEATURES

The new removable heated glass build plate and upgraded extruder nozzle allow for printing in ABS and nylon. A new cloud slicer lets you remotely access your printer from anywhere with your Dremel account, and its built-in "fix" option will try to fill errors in your model before directing you to print settings. The slicer software works well but could definitely use some tweaks. It's a brand-new platform, so we'll give it the benefit of the doubt that they are actively improving it.

BUILT FOR EDUCATORS

Dremel actively markets this machine as a learning device and their design embodies that. It produces quality, repeatable results in a compact, fully enclosed, quiet package. ◉

WHY TO BUY
Sensors everywhere! Open doors, filament failure, and climate control sensors make this machine safe to operate and less wasteful.

- **WEBSITE** 3dprinter.dremel.com
- **MANUFACTURER** Dremel
- **BUILD VOLUME** 255×155×170mm
- **BED STYLE** Heated with removable glass
- **FILAMENT SIZE** 1.75mm
- **OPEN FILAMENT?** Yes, but chipped filament is optimal
- **TEMPERATURE CONTROL?** Yes, extruder (280°C max); bed (100°C max)
- **PRINT UNTETHERED?** Yes, Wi-Fi, SD card
- **ONBOARD CONTROLS?** Yes, full color touchscreen
- **HOST/SLICER SOFTWARE** Dremel IdeaBuilder
- **OS** Mac, Windows
- **FIRMWARE** Custom, proprietary
- **OPEN SOFTWARE?** No
- **OPEN HARDWARE?** No

MONOPRICE SELECT MINI V2

A capable machine that won't break the bank   Written by Adam Casto

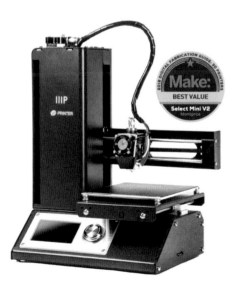

2018 DIGITAL FABRICATION GUIDE 3D PRINTERS
Make:
BEST VALUE
Select Mini V2
Monoprice

	0	1	2	3	4	5
VERTICAL SURFACE FINISH						
HORIZONTAL FINISH						
DIMENSIONAL ACCURACY						
OVERHANGS						
BRIDGING						
NEGATIVE SPACE						
RETRACTION						
SUPPORT MATERIAL						
SQUARENESS						
FULL BED ACCURACY						
Z WOBBLE		PASS				

MACHINE RATING 41
Price as Tested **$220**

YOU MIGHT BE SKEPTICAL AT ITS $220 PRICE POINT, but the diminutive Monoprice Select Mini V2 delivers. With its small but usable build volume and impressive feature set it brings tremendous utility to the table.

A PLEASANT SURPRISE

The steel body is unexpectedly sturdy, and the heated bed is insulated with a textured build surface. It also sports a simple 3.7" display with a functional knob-controlled interface. You can print from the microSD slot, USB, or even Wi-Fi. There's also a remarkably robust online community.

LIMITATIONS, NOT DEAL-BREAKERS

Printing ABS can be a challenge since the bed temperature maxes out at 60°C, and proprietary firmware and onboard controls may be a sticking point, although there are workarounds. Whether you're looking for a first printer or a secondary standby, you can't go wrong with the Select Mini V2. ◉

WHY TO BUY
Whether you're just getting started or an experienced user looking for a little versatile printer, the Select Mini V2 offers performance at a price point that can't be overlooked.

- **WEBSITE** monoprice.com
- **MANUFACTURER** Monoprice
- **BUILD VOLUME** 120×120×120mm
- **BED STYLE** Heated, insulated with textured build surface
- **FILAMENT SIZE** 1.75mm
- **OPEN FILAMENT?** Yes
- **TEMPERATURE CONTROL?** Yes, extruder (250°C max); bed (60°C max)
- **PRINT UNTETHERED?** Yes, Wi-Fi, microSD card
- **ONBOARD CONTROLS?** Yes, 3.7" color IPS display
- **HOST/SLICER SOFTWARE** Cura
- **OS** Mac, Windows, Linux
- **FIRMWARE** Proprietary
- **OPEN SOFTWARE?** Yes, Cura is AGPLv3
- **OPEN HARDWARE?** No

Hep Svadja

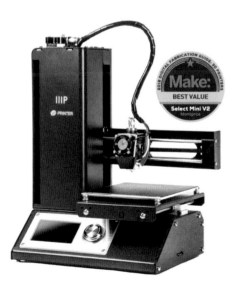

TAZ 6 Written by Darius McCoy

This large, tried-and-true machine is an open source dream

THE TAZ 6 FEATURES PRETTY MUCH EVERYTHING YOU'D WANT IN A PRINTER with a large build volume, auto-leveling, auto-nozzle cleaning, and tetherless printing.

AT LONG LAST: AUTO-LEVELING

LulzBot finally incorporated auto-leveling, completed by using conductive sensing, in both of their flagship printers, the Mini and Taz 6. Another addition is LulzBot's new Tool Head v2.1. Only available for the Taz 6 right now, it includes a new extruder body and a bigger cooling fan that LulzBot says will prolong the life of the hot end.

THE BIG EASY

Overall, LulzBot created a big and reliable printer with top of the line features that gives it broad appeal. Don't let the $2,500 price scare you away; if you have the budget for it you won't regret buying one. ⦿

WHY TO BUY
We consider it to be one of the best open source printers out on the market in both terms of structural design and printing quality.

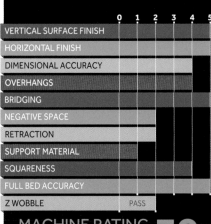

	0	1	2	3	4	5
VERTICAL SURFACE FINISH						
HORIZONTAL FINISH						
DIMENSIONAL ACCURACY						
OVERHANGS						
BRIDGING						
NEGATIVE SPACE						
RETRACTION						
SUPPORT MATERIAL						
SQUARENESS						
FULL BED ACCURACY						
Z WOBBLE	PASS					

MACHINE RATING 39
Price as Tested **$2,500**

- **WEBSITE** lulzbot.com
- **MANUFACTURER** LulzBot
- **BUILD VOLUME** 280×280×250mm
- **BED STYLE** Heated with PEI surface
- **FILAMENT SIZE** 2.85mm
- **OPEN FILAMENT?** Yes
- **TEMPERATURE CONTROL?** Yes, extruder (300°C max); bed 120°C max)
- **PRINT UNTETHERED?** Yes, SD card
- **ONBOARD CONTROLS?** Yes, scroll knob and LCD
- **HOST/SLICER SOFTWARE** Cura LulzBot Edition
- **OS** Mac, Windows
- **FIRMWARE** Marlin
- **OPEN SOFTWARE?** Yes, GPLv3
- **OPEN HARDWARE?** Yes, GPLv3

ZORTRAX M300 Written by Jonathan Prozzi

Great for a professional studio that needs a solid printer

THE ZORTRAX M300 BOASTS A LARGE BUILD VOLUME, PRECISE PRINTS, and support for a variety of materials. The assisted calibration is smooth and accurate, giving feedback as you tighten screws underneath the bed.

PLUSES AND MINUSES

One standout feature is the perforated bed, which resulted in extremely solid first layers. The mandatory rafts were easy to remove, but you can't cancel while they're printing. I found it lacked some features that I'd expect at this price point, with just a single extruder, a long ramp-up time for each print, and a small and sometimes hard to read display screen.

ACCURATE, HIGH-QUALITY PRINTS

The Z-Suite software is great for those who don't want to tinker: just install, load a design, set the material, and slice. The M300 could be good for producing large, low-maintenance yet extremely high-quality prints. ⦿

WHY TO BUY
The M300 is a sturdy, reliable printer that produces precise prints without the need to get under the hood or adjust settings.

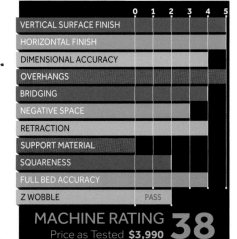

	0	1	2	3	4	5
VERTICAL SURFACE FINISH						
HORIZONTAL FINISH						
DIMENSIONAL ACCURACY						
OVERHANGS						
BRIDGING						
NEGATIVE SPACE						
RETRACTION						
SUPPORT MATERIAL						
SQUARENESS						
FULL BED ACCURACY						
Z WOBBLE	PASS					

MACHINE RATING 38
Price as Tested **$3,990**

- **WEBSITE** zortrax.com
- **MANUFACTURER** Zortrax
- **BUILD VOLUME** 300×300×300mm
- **BED STYLE** Heated, perforated surface
- **FILAMENT SIZE** 1.75mm
- **OPEN FILAMENT?** Yes, as of April 2017
- **TEMPERATURE CONTROL?** No for Zortrax brand filaments, Yes for third party, extruder (380°C max); bed (110°C max)
- **PRINT UNTETHERED?** Yes, SD card
- **ONBOARD CONTROLS?** Yes, LCD screen with analog wheel for selections
- **HOST/SLICER SOFTWARE** Z-Suite
- **OS** Mac OS X, Windows 7 (or later)
- **FIRMWARE** Z-Firmware
- **OPEN SOFTWARE?** No
- **OPEN HARDWARE?** No

Hep Svadja

LULZBOT MINI
Written by Adam Casto

Quality and simplicity make this a solid choice for any setting

	0	1	2	3	4	5
VERTICAL SURFACE FINISH						
HORIZONTAL FINISH						
DIMENSIONAL ACCURACY						
OVERHANGS						
BRIDGING						
NEGATIVE SPACE						
RETRACTION						
SUPPORT MATERIAL						
SQUARENESS						
FULL BED ACCURACY						
Z WOBBLE		PASS				

MACHINE RATING 37
Price as Tested **$1,250**

IF YOU NEED A STURDY AND RELIABLE PRINTER you can't go wrong with the Mini.

LESS IS MORE
Sporting only a power switch and USB port on the front panel, the LulzBot Mini lacks many of the bells and whistles compared to other printers at this price point. To some this may be a limitation, but simplicity is what makes this a wonderful printer.

ATTENTION TO DETAIL
The LulzBot line's quality is obvious in things like the sturdy steel frame, beefy lead screws, and geared extruder with an all-metal hot end capable of handling almost any filament. Finer points include self-lubricating polymer bushings and dampers on stepper motors. The default settings in the easy-to-setup LulzBot edition of Cura consistently turn out good prints. The Mini will fit in nicely anywhere, with both new and advanced users alike. ◐

WHY TO BUY
Reliability and ease of use make this rugged little workhorse a good choice for the home, shop, classroom, or makerspace.

- **WEBSITE** lulzbot.com
- **MANUFACTURER** LulzBot
- **BUILD VOLUME** 152×152×158mm
- **BED STYLE** Heated glass with PEI print surface
- **FILAMENT SIZE** 2.85mm
- **OPEN FILAMENT?** Yes
- **TEMPERATURE CONTROL?** Yes, extruder (300°C max); bed (120°C max)
- **PRINT UNTETHERED?** No
- **ONBOARD CONTROLS?** No
- **HOST/SLICER SOFTWARE** Cura LulzBot Edition
- **OS** Mac, Windows, Linux
- **FIRMWARE** Marlin-based
- **OPEN SOFTWARE?** Yes, Cura is AGPLv3
- **OPEN HARDWARE?** Yes, GPLv3 and CC-BY-SA 4.0

PRINTRBOT SMALLS LIMITED EDITION
Written by Ryan Priore

Without a hiccup during our testing, it just kept cranking out prints

	0	1	2	3	4	5
VERTICAL SURFACE FINISH						
HORIZONTAL FINISH						
DIMENSIONAL ACCURACY						
OVERHANGS						
BRIDGING						
NEGATIVE SPACE						
RETRACTION						
SUPPORT MATERIAL						
SQUARENESS						
FULL BED ACCURACY						
Z WOBBLE		PASS				

MACHINE RATING 37
Price as Tested **$500**

ALTHOUGH THE SMALLS MAY LOOK LIKE THE PRINTRBOT SIMPLE'S LITTLE BROTHER, it boasts a larger print volume, a smaller overall footprint, and perhaps even a little more personality and panache.

LIMITED VS. STANDARD
The Limited Edition features a slightly larger build platform, linear rails over the standard's linear rods, and an attractive wood trim exterior. At $500 it's also pricey compared to the $398 standard and $298 kit and doesn't come with a heated bed.

A RELIABLE MACHINE
The prints looked fantastic, and the auto-tramming operation led to a solid "set and forget" print process. A Cura profile is supplied for the Smalls, but I needed to manually edit the machine settings in order to achieve the full benefits of the build area. The stock speeds also seemed a bit slow, but I'll take quality prints over quickness. ◐

WHY TO BUY
Die-hard Printrbot fans will rejoice at the Smalls aesthetics, form factor, and overall simplicity, but the Limited Edition price may be a stretch over the standard Smalls.

- **WEBSITE** printrbot.com
- **MANUFACTURER** Printrbot
- **BUILD VOLUME** 173×150×148mm
- **BED STYLE** Non-heated aluminum print surface
- **FILAMENT SIZE** 1.75mm
- **OPEN FILAMENT?** Yes
- **TEMPERATURE CONTROL?** Yes, extruder (270°C max)
- **PRINT UNTETHERED?** Yes, SD card
- **ONBOARD CONTROLS?** No
- **HOST/SLICER SOFTWARE** Cura
- **OS** Mac, Windows, Linux
- **FIRMWARE** Marlin
- **OPEN SOFTWARE?** Yes, Cura is AGPLv3
- **OPEN HARDWARE?** Yes, CC-BY-SA 3.0

Hep Svadja

VERTEX NANO
Written by Kelly Egan

This machine's biggest appeal is its small size

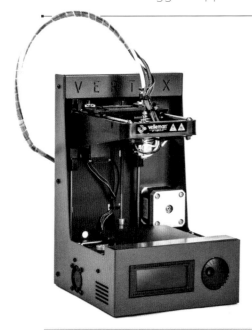

THE VERTEX NANO IS TINY, WITH A 3" SQUARE BED, BUT THERE ARE PLENTY of projects you could print on it. Print quality is generally good, with no problems using the default settings. The print surface is unheated BuildTak, so you'll likely want to stick to PLA.

SOME QUIRKS

The recommended slicer is currently only available for Windows. You can use another, but make sure your endstops are set up correctly. There's no bed leveling, no extruder fan, and the printer doesn't move away from a finished print.

TAKE IT TO GO

Aside from the fan, most problems could be fixed in firmware updates. We tested an assembled version — if you're building your own, it may take some time to calibrate correctly. Still, a printer you can literally fit in your toolbox seems like a great option. ✪

WHY TO BUY
The Vertex Nano is a tiny printer that would work well in cramped quarters. It's small enough you could take it with you.

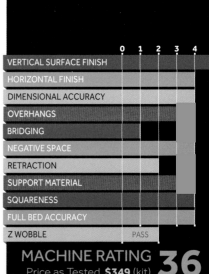

Category	0	1	2	3	4	5
VERTICAL SURFACE FINISH						
HORIZONTAL FINISH						
DIMENSIONAL ACCURACY						
OVERHANGS						
BRIDGING						
NEGATIVE SPACE						
RETRACTION						
SUPPORT MATERIAL						
SQUARENESS						
FULL BED ACCURACY						
Z WOBBLE	PASS					

MACHINE RATING 36
Price as Tested **$349** (kit)

- **WEBSITE** vertex3dprinter.eu
- **MANUFACTURER** Velleman
- **BUILD VOLUME** 80×80×75mm
- **BED STYLE** Non-heated with BuildTak
- **FILAMENT SIZE** 1.75mm
- **OPEN FILAMENT?** Yes
- **TEMPERATURE CONTROL?** Yes, extruder (245°C max)
- **PRINT UNTETHERED?** Yes, SD card
- **ONBOARD CONTROLS?** Yes, LCD display, analog wheel selections
- **HOST/SLICER SOFTWARE** Vertex Nano Repetier-Host
- **OS** Available for Windows only (or use another slicer compatible with your OS)
- **FIRMWARE** Marlin
- **OPEN SOFTWARE?** Yes, Repetier is Apache-2.0
- **OPEN HARDWARE?** No

MAKEIT PRO-L
Written by Chris Yohe

Build a batch-printing farm with this sizeable machine

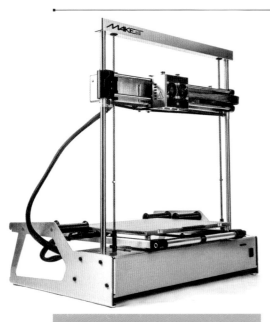

AIMED AT A SLIGHTLY HIGHER MARKET, the MakeIt Pro-L provided decent print quality with a large surface area.

NOVEL FEATURES

At the back of the bed is a great nozzle cleaning wipe pad that helped during printing. The Pro-L also touts a novel "duplication" feature, taking advantage of the dual extruders for small prints.

PREMIUM PRICE, AVERAGE RESULTS

The printer itself functioned fine, but the high price was not met with high-quality results. The included documentation was good for troubleshooting feeding and clogging issues with the extruders and hot ends, but there were a few too many for our taste. Some cooling enhancements could help, but the biggest hurdle is the price. If you're putting together a print farm, the Pro-L may fulfill your batch printing needs, but be ready to get your hands dirty. ✪

WHY TO BUY
A large, sturdy build platform paired with dual extruders and utilitarian design makes this a workhorse for professional users.

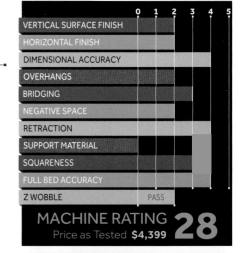

Category	0	1	2	3	4	5
VERTICAL SURFACE FINISH						
HORIZONTAL FINISH						
DIMENSIONAL ACCURACY						
OVERHANGS						
BRIDGING						
NEGATIVE SPACE						
RETRACTION						
SUPPORT MATERIAL						
SQUARENESS						
FULL BED ACCURACY						
Z WOBBLE	PASS					

MACHINE RATING 28
Price as Tested **$4,399**

- **WEBSITE** makeit-3d.com
- **MANUFACTURER** MakeIt
- **BUILD VOLUME** 305×254×330mm (single and dual extrusion)
- **BED STYLE** Heated aluminum; magnetic removable steel sheet with Ultralex
- **FILAMENT SIZE** 1.75mm
- **OPEN FILAMENT?** Yes
- **TEMPERATURE CONTROL?** Yes, extruder (2__ max); bed (120°C max)
- **PRINT UNTETHERED?** Yes, USB, S__
- **ONBOARD CONTROLS?** Yes, LCD screen and scroll wheel
- **HOST/SLICER SOFTWARE** Cura, MatterControl, Simplify3D
- **OS** Mac, Windows, Linux
- **FIRMWARE** Custom
- **OPEN SOFTWARE?** No
- **OPEN HARDWARE?** No

Hep Svadja

ShopBot CNC Puts Your Ideas Into Production

Everyone has ideas and with a ShopBot CNC, you have the power to make them a reality. Go from prototype to production all with the same tool. Whatever size your project is, there's a ShopBot to help you make it.

Once you get through the prototype phase of your project, you might think that you need to have a small run done by a large scale manufacturer to test out the final product. Not necessarily, if you have a ShopBot CNC tool. An example? You don't have to look any further than the pop-up factory we created at World Maker Faire New York 2017 for our Portable Picnic Trays.

Designed and beta tested before going to World Maker Faire, the tray is made from a couple of parts—all of which are cut on a ShopBot Desktop. In true maker fashion, we hand assembled the trays on-site in New York—taking them through our pop-up factory production and assembly line to arrive at the finished product. You can read more about the development of the trays on the ShopBot blog at www.ShopBotBlog.com

Also on the ShopBot blog are stories of people using ShopBot CNC tools to create all kinds of products. Everything from furniture to sports novelties to gaming accessories to musical instruments. Often, what starts as a singular idea grows exponentially and winds up moving from the realm of hobby to the world of business.

Getting your product started doesn't have to require a lot of people or the backing of a corporation. *As long as you have your idea and your ShopBot, anything is possible!*

Discover our full line of tools on our website. **Then give us a call.** We'll help you choose the right tool to bring your ideas to life.

We make the tools for making the future.

888-680-4466 • ShopBotTools.com

Photography by Sarah E. Gibbons

XFAB

Stellar prints make this a worthy competitor to the Formlabs empire Written by Matt Stultz

THE XFAB FROM ITALIAN MANUFACTURER DWS IS ONE OF THE PRICIER PRINTERS WE'VE TESTED over the years, but also one of the more capable. With a large build volume, wide range of materials, and incredible print quality, the XFab is impressive.

BIGGER AND BETTER

One of the downsides to many SLA printers is the limited size. You can have pretty, just not a lot of it. However, the XFab is big, and the grooved 180mm circular plate provides plenty of build space. Prints stick well to the surface, and the grooves make the prints easy to remove, something I can't say for other SLA machines.

More than 11 different resin formulations are available for the XFab, which come in injection tubes that are squeezed like a caulking gun to automatically fill the vat. This is a no-mess method of getting the resin into the machine, but I wish it were smart enough to only fill the amount needed for the job at hand. The leftover resin separated over time and needed to be stirred before the next use.

CLOSED-SOURCE HURDLES

My biggest issue with the XFab is that it's just too closed. If you don't have the resin cartridges in the machine correctly, you get an error because the machine can't read their RFID tags. If your internet connection drops, the software won't let you operate the machine because it can't authenticate the software license with the server. For $6K, users should be able to run their printer without worrying about spotty internet.

EXCEPTIONAL PRINTS

That said, my prints were some of the best I've seen on any SLA machine. The XFab is a true SLA, using lasers to cure the UV-sensitive resin. While you can often see layer lines more dramatically in SLA prints than you can from DLP, the XFab prints were very clean.

If you're looking for a pro level SLA, the XFab clearly shows there are other options out there beyond the Formlabs machines that dominate the pro-desktop market. For $6,000 though, I don't see a lot of hobbyists jumping into these Italian waters. ◐

Price as Tested $6,000

■ **WEBSITE** www.dwslab.com	■ **HOST/SLICER SOFTWARE** Nauta XFab Edition and Fictor XFab Edition
■ **MANUFACTURER** DWS Lab	
■ **BUILD VOLUME** 180mm (dia.)×180mm (height)	■ **OS** Windows XP, 7, 8 (32- and 64-bit)
■ **OPEN RESIN?** No, chipped	■ **FIRMWARE** Proprietary
■ **PRINT UNTETHERED?** No	■ **OPEN SOFTWARE?** No
■ **ONBOARD CONTROLS?** No	■ **OPEN HARDWARE?** No

PRO TIPS
Plan out your work area to minimize the space between your laptop and the machine. When you unlock the lid, it doesn't stay that way very long and can only be unlocked by clicking a button on your computer in their software.

WHY TO BUY
The DWS XFab creates fantastic prints and has a wide range of resins available. If you want options in your SLA without giving up quality, this is a great printer for you.

TEST PRINT

Hep Svadja

THE XFAB CLEARLY SHOWS THERE ARE OTHER OPTIONS OUT THERE BEYOND THE FORMLABS MACHINES

XFAB

DUPLICATOR 7

An economical, high-quality resin printer **Written by Matt Stultz**

CHINESE MANUFACTURER WANHAO HAS RELEASED THE DUPLICATOR 7, often called D7 — a Liquid Crystal Display-based resin printer that provides great results with an equally great price tag.

A NEW LIGHT

A series of UV LEDs are mounted under a normal LCD. When the pixels are black, light can't pass through, but set to white (clear really) it does, curing the resin. My test prints were clean, and fine details were crisp — on par with any DLP or SLA prints from other consumer machines. So while LCD may be new, it works well. The D7 is still a work in progress, though, and the company has been keeping up with community requests for mods to make it better.

PRICE THAT RESONATES

At a cost attractive to new users, the D7 is going to really expand the adoption of resin printing. If you are not ready to jump in with

Base price **$495**

■ **WEBSITE** wanhao3dprinter.com	■ **HOST/SLICER SOFTWARE** Creation Workshop
■ **MANUFACTURER** Wanhao	■ **OS** Windows
■ **BUILD VOLUME** 120×68×200mm	■ **FIRMWARE** Proprietary
■ **OPEN RESIN?** Yes	■ **OPEN SOFTWARE?** No
■ **PRINT UNTETHERED?** No	■ **OPEN HARDWARE?** No
■ **ONBOARD CONTROLS?** No	

PRO TIPS
The community has configured the Raspberry Pi-powered NanoDLP package to work with the D7 — you no longer need to be connected to a full-blown Windows computer to operate.

WHY TO BUY
The D7 is a high-quality resin printer at an attractive price point, and the easily reproducible vat makes it a great fit for workshops and makerspaces with multiple users.

an expensive SLA, the Duplicator 7 is a great place to get started. ◒

MOAI

Use Cura and print untethered with this impressive SLA machine **Written by Matt Stultz**

WHILE THERE HAVE BEEN NUMEROUS KIT OPTIONS FOR DLP-BASED RESIN PRINTERS, the Moai is the first I know to be a galvanometer-based SLA. Building the kit was quick — we were able to get it fabricated and calibrated in an afternoon.

QUALITY OUTPUT

Using one of Peoploy's three provided profiles, test prints came out clean with no failures. Even though I prefer the quality of DLP-based prints, the Moai prints were more than acceptable for being laser-based.

EASY AND UNTETHERED

In a market full of proprietary software and tethered machines, the Moai uses Cura and prints from SD card. I can't stress enough how much easier that makes using this machine. The community contributions to Cura have made it very user friendly and printing from an SD card thankfully frees up

Price as Tested **$1,250**

■ **WEBSITE** peopoly.net	■ **ONBOARD CONTROLS?** Yes, pushbutton scroll knob and LCD display
■ **MANUFACTURER** Peopoly	■ **HOST/SLICER SOFTWARE** Cura Moai Edition
■ **BUILD VOLUME** 130×130×180mm	■ **OS** Mac OSX, Windows
■ **OPEN RESIN?** Yes	■ **FIRMWARE** Proprietary
■ **PRINT UNTETHERED?** Yes, SD card	■ **OPEN SOFTWARE?** No
	■ **OPEN HARDWARE?** No

PRO TIPS
The kit includes a pair of laser filtering safety glasses that need to be worn during the calibration process — wear them! It is powerful enough to cause eye damage.

WHY TO BUY
Finding a laser-based SLA that isn't tied to proprietary resins is a hard thing to do; finding one that also costs half of most of its competitors is amazing.

Hep Svadja

your computer.

I was really impressed by the Moai. It provides great prints and with much less hassle than many other resin printers. ◒

BENCHTOP PRO

A compact and customizable production machine   Written by Matt Dauray

THE BENCHTOP PRO FROM CNC ROUTER PARTS IS A SMALLER VERSION of their popular Pro line. It takes all the beef and brawn of the bigger setups and shortens them up to fit in a 41"×41" machine footprint (with motors attached). The all-metal construction still accepts a range of router options, from smaller HP off-the-shelf routers to its own 3HP integrated spindle. This particular model had the DeWalt 2¼HP router which further adds to utility, as it runs on 110V. We've tested the larger versions in the past, and the features that made them great carry on to this model. Being able to buy modules to fit your needs makes this a truly customizable machine, while still staying in a competitive price range.

OPTIMIZED FOR OUTPUT

CNC Router Parts creates machines capable of running in a production environment — designed to be used heavily and with repeatability. Special attention is paid to things like cable management, ensuring that continued use is not interrupted by annoying problems like strained wires. The compact build space lends itself to a maker or business needing high output for smaller products.

ADVANCED SOFTWARE

The software recommended by CNC Router Parts, Mach3, takes no prisoners in terms of user friendliness. Mach3 is a fully featured machine control software that works wonderfully, provided you know what all the little buttons and doodads do, but it's very easy to break bits and ruin parts if you are a button masher. Keep the e-stop handy and an eye on the machine. Someone who knows their way around machine control software will be right at home — there's all sorts of bells and whistles to make your cuts faster, more precise and generally more efficient.

SMALL BUT MIGHTY

The Benchtop Pro is a great solution for a maker or business with space constraints but a need for production quality. Considering that most of its parts are interchangeable with its bigger brothers, whatever needs you might have in your small space can almost certainly be achieved. ◢

Base price **$3,250**

PRO TIPS

Use practice runs to dial in your programs. Load the G-code and run it with no bit and the router off, with the z set high. Look carefully at your preview to make sure all your paths are there. If you're using F360, make sure the entire CAM setup is selected when you post, because it is possible to post individual paths. Trust me.

WHY TO BUY

Shop space is at a premium these days, but no one wants to give up quality in their pursuit of compact efficiency. The Benchtop Pro provides all the rigidity and precision of its bigger brothers in a compact package. If you buy the smaller base and DeWalt router, you can run on 110V instead of 220V with the integrated spindle.

- **WEBSITE**
  cncrouterparts.com
- **MANUFACTURER**
  CNC Router Parts
- **PRICE AS TESTED**
  $4,062
- **ACCESSORIES INCLUDED AT BASE PRICE**
  None
- **ADDITIONAL ACCESSORIES PROVIDED FOR TESTING**
  Wasteboard, router mount, 2¼HP router, electronics package, proximity switches, auto z and corner, Mach3, ¼" and ½" bits
- **BUILD VOLUME**
  635×635×171mm
- **MATERIALS HANDLED**
  Woods, plastics, soft metals
- **WORK UNTETHERED?**
  No, Ethernet connection to computer running Mach3
- **ONBOARD CONTROLS?**
  Yes, control box master switch, separate motors switch, e-stop, power switch on router
- **DESIGN SOFTWARE**
  Fusion 360 recommended
- **CUTTING SOFTWARE**
  Mach3, VCarve, Aspire, Cut2D, Cut3D
- **OS**
  Depends on cutting software
- **FIRMWARE**
  Custom proprietary
- **OPEN SOFTWARE?**
  No
- **OPEN HARDWARE?**
  No

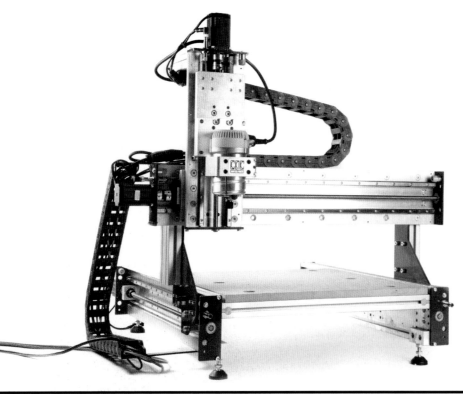

Hep Svadja

THE COMPACT BUILD SPACE LENDS ITSELF TO A MAKER OR BUSINESS NEEDING HIGH OUTPUT FOR SMALLER PRODUCTS

HANDIBOT 2
Written by Matt Stultz

ShopBot rethinks portability with this unique design

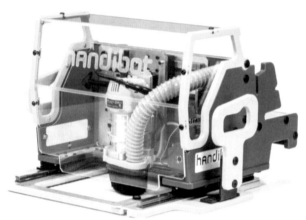

HANDIBOT 2 DOESN'T WORK QUITE LIKE A NORMAL CNC ROUTER. Instead, it sits on top of your workpiece and can be repositioned around the item it's carving, which means the size of the item you want to cut is only limited by the number of times you want to move the Handibot.

CRISP CUTS

Cutting on the Handibot 2 is easy and effective. Its router is the standard DeWalt 611 used in many CNC machines. At 1¼HP, it provides plenty of power and cuts come out clean. An Intel Edison onboard provides for untethered carving. There's also a protective cover and a hose to hook up a shop vac for dust and debris.

PORTABLE POWERHOUSE

While I don't think the Handibot 2 is a complete replacement for a large CNC, there's a lot of power packed in that little package, and for those looking for an on-the-go CNC, the Handibot 2 is a little titan. ◔

Base price **$3,195**

- **WEBSITE** shopbottools.com
- **MANUFACTURER** ShopBot
- **PRICE AS TESTED** $3,195
- **ACCESSORIES INCLUDED AT BASE PRICE** LMT Onsrud 37-61 ½" 90° V-bit, 61-040 ⅛" O flute straight-cut, and 52-287 ¼" 2 flute up-cut, DeWalt DWP 611 router
- **ADDITIONAL ACCESSORIES PROVIDED FOR TESTING** None
- **BUILD VOLUME** 152×203×76mm
- **MATERIALS HANDLED** Wood, plastics, soft metals
- **WORK UNTETHERED?** Yes, wireless
- **ONBOARD CONTROLS?** Yes, two control buttons
- **DESIGN SOFTWARE** VCarve Pro ShopBot Edition
- **CUTTING SOFTWARE** FabMo
- **OS** Windows for VCarve; FabMo is onboard the machine and can be accessed via any web browser
- **FIRMWARE** FabMo
- **OPEN SOFTWARE?** Yes, FabMo is Apache-2.0
- **OPEN HARDWARE?** Yes, Panaka OHL, Version 1.0

PRO TIPS
Building jigs for the Handibot 2 makes it much easier for you to complete repetitive small jobs, and the base of the Handibot can easily lock into those jigs.

WHY TO BUY
The Handibot 2 puts a lot of power in a small package. If you want a portable CNC that punches above its weight class, the Handibot 2 is up for it.

CNC-STEP HIGH-Z S400T

This versatile machine is built to last
Written by Simon Norridge

THE BENCHTOP CNC-STEP HIGH-Z S400T OFFERS A GREAT BUILD QUALITY and plenty of tools to keep you busy. We could have lumped this machine in with the Hybrids thanks to its swappable tool system, but this machine really shines as a CNC milling and engraving device.

SUPER SOFTWARE

Wiring up the controller and engraving head was easy with the aid of the excellent documentation that came with the machine.

The KinetiC-NC control software has a well designed UI that makes controlling the machine very accurate and easy to perform. It's the best CNC UI I've used and I'm so looking forward to working with it more. The Otto Suhner UAD 30 RF Milling/Engraving router is also robust and simple to operate.

PERFECT PERFORMANCE

My first introduction to engraving machines was over 40 years ago, and

Base price **$5,299**

- **WEBSITE** cncstepusa.com
- **MANUFACTURER** CNC-Step
- **PRICE AS TESTED** $14,730
- **ACCESSORIES INCLUDED AT BASE PRICE** Zero3 5-channel controller, Suhner UAD 30-RF router spindle (3,500–30,000rpm, 1050W), WinPC-NC USB v3.0 controller software
- **ADDITIONAL ACCESSORIES PROVIDED FOR TESTING** Laser engraver, cooling system, diamond engraver, tangential cutter, 4th axis rotary table
- **BUILD VOLUME** 400×300×110mm
- **MATERIALS HANDLED** Wood, plastic, non-ferrous metals, stone, and ferrous metals with the right tools/speeds/feeds
- **WORK UNTETHERED?** No
- **ONBOARD CONTROLS?** No, e-stop only
- **DESIGN SOFTWARE** Any CAD package should work fine
- **CUTTING SOFTWARE** KinetiC-NC
- **OS** Windows 8–10
- **FIRMWARE** FabMo
- **OPEN SOFTWARE?** No
- **OPEN HARDWARE?** No

PRO TIPS
The software thumb drive is also a bottle opener!

WHY TO BUY
A solid CNC machine in its own right, the swappable tool system and great documentation make the High-Z S400T an attractive machine.

I'm confident in stating the CNC-Step HIGH-Z S400T outperformed any such similar machine I've used in the past. ◔

Hep Svadja

GLOWFORGE BASIC
Written by Matt Stultz

The feature-rich, easy-to-use machine is finally shipping

WE HAD THE GLOWFORGE PLUGGED IN, SET UP, AND MAKING CUTS within 10 minutes. Its only data connection is via Wi-Fi, and the web app ensures software is always up to date — we've been pleased with the improvements we've seen already. Glowforge accepts a number of image files for etching, but the mainstay is vector graphics; it handles Inkscape-created SVG files perfectly.

PERFORMANCE
Our first cuts were clean and easy. The built-in camera and the software work together to make the Glowforge the easiest digital fabrication machine we've used. We experienced timing issues on the pre-release machine reviewed in *Make:* Vol. 56, but the production units are cutting and engraving just fine.

NOW AVAILABLE
The official launch was fraught with delays, but worth the wait. Preorders have been shipping since this summer. Meanwhile, the optional external air filter is slated to ship in December ... maybe. ⊘

Base price $2,995

PRO TIPS
Use color mapping on your SVG files to change the order and type of operation on parts of your design.

Make sure to follow the guidelines for venting — the built-in fan can do its job, but not if you make it work too hard. As with any laser, an air filtration unit isn't a bad idea.

WHY TO BUY
The Glowforge is an extremely easy-to-use laser cutter that takes away many of the software pain points that plague other machines.

- **WEBSITE** glowforge.com
- **MANUFACTURER** Glowforge
- **ACCESSORIES INCLUDED AT BASE PRICE** 40W laser tube, 6 mo. warranty
- **BUILD VOLUME** 290×515mm
- **TUBE** 40W
- **CUT UNTETHERED?** Yes, Wi-Fi
- **MATERIALS HANDLED** Cutting and engraving: wood, fabric, leather, paper, acrylic, acetal, mylar, rubber, cork; engraving only: glass, marble, stone, ceramic tile, anodized aluminum, titanium
- **ONBOARD CONTROLS?** Yes, single control button
- **SOFTWARE** Glowforge web interface, no installed software
- **OS** Mac, Windows, Linux
- **FIRMWARE** Proprietary
- **OPEN SOFTWARE?** No
- **OPEN HARDWARE?** No

FULL SPECTRUM MUSE
Written by Jen Schachter

A sleek new challenger in the desktop hobby laser arena

FULL SPECTRUM'S 6TH-GEN MACHINE, THE MUSE FAVORS BEGINNERS, with hyper-useful features like an onboard camera for alignment, touchscreen controls, and foolproof warning systems, although the manual z-height focus is incongruously analog.

NEW TRICKS, OLD DOG
Full Spectrum shares their years of laser experience through free tutorials, sample projects, and an e-book available on their site. It's a gold mine of information for laser newbies. There is also an active forum community, but it took persistence to reach anyone in customer support for troubleshooting.

RetinaEngrave v2 (RE2) is Muse's proprietary control software, which connects via Ethernet or Wi-Fi. Despite a streamlined interface, RE2 can be finicky for advanced operations and importing complex files from external design programs. Etching and photo engraving worked well, but we struggled with the software. ⊘

Base price $5,000

PRO TIPS
A "material test" file and speed/power/current table is available at laser101.fslaser.com/materialtest. Use that as a starting point and test in both directions until you get the desired effect. Record your results, or better, make sample tiles to show a range of setting options for each material.

- **WEBSITE** fslaser.com
- **MANUFACTURER** Full Spectrum Laser
- **ACCESSORIES INCLUDED AT BASE PRICE** 40W laser tube, basic water pump, 1 yr. warranty
- **BUILD VOLUME** 508×305mm
- **TUBE** 40W (upgradable to 45W)
- **CUT UNTETHERED:** Yes, Wi-Fi
- **MATERIALS HANDLED** Cutting and engraving: wood, acrylic, fiber, leather, paper and cardboard; engraving only: metal, glass, and curved items
- **ONBOARD**

STEPCRAFT 2/840
Written by Simon Norridge

This capable machine has just about every tool head you could need

THE STEPCRAFT 840 ISN'T THE FIRST CNC/3D PRINTER HYBRID WE'VE LOOKED at, but it might be the most capable. With a variety of tools available and a large work area, there are few tasks you can't complete with the 840. Absorb the manual well before you start the kit assembly process, and be patient and methodical. You can easily make mistakes with the wiring and cable routing that require stripping down and reassembly.

IMPRESSIVE OPTIONS

CNC machining with this machine was so quiet, so smooth, and so easy to control. At first the Stepcraft 840 looks all CNC router-ish. Then you discover a 3D printer head that is a breeze to fit onto the machine. I started with printing the small Stepcraft demo logo. It took longer than I had hoped to get the filament feeding well. Maybe a few modifications to the drive wheel and a torque setting will help to alleviate those issues. It 3D prints, but needs some tuning to run optimally.

The 840 has additional options for laser etching, wood burning, hot-wire cutting, pen-plotting, and more. The scribe is another easily attached tool. Well-made and spring loaded, it appears to have the strength and durability to scribe many different materials. I used cast 6mm thick acrylic and it worked well.

SUGGESTED UPGRADES

One concern is the strength and durability of the machine bed, which is laminated MDF board. Using counter-bored cap head screws in place of the socket flanged button screws would give a greater degree of rigidity. Some users may want to consider drilling and dowel-pinning a few of the assembled parts to ensure continued alignment. I had to undo, re-adjust, and re-tighten the machine to ensure smooth motion on a couple of occasions.

UTILITY PLAYER

Like other hybrid machines, the Stepcraft 840 may not be a master of any of the tools it brings to the table, but with more time to tune it in it could be a "swiss army knife" for makers. ◉

Base price $2,199

PRO TIPS
You may want to consider drilling and dowel pinning a few of the assembled parts to ensure continued alignment.

Read and absorb the manual well before you start the assembly process. Mistakes can easily be made with the wiring and cable routing that result in the need to strip down and reassemble.

WHY TO BUY
With the 840's large work area and wide variety of tools, there are few tasks you can't complete.

- **WEBSITE** stepcraft.us
- **MANUFACTURER** Stepcraft
- **TOOL HEADS AVAILABLE** Milling, 3D printing, engraving, laser etcher, wood burner, hot wire, drag knife, plotting pen, and more
- **BUILD VOLUME** 600×840×140mm
- **WORK UNTETHERED?** No
- **ONBOARD CONTROLS?** E-stop only
- **CONTROL SOFTWARE** UCCNC
- **OS** Windows
- **FIRMWARE** Proprietary
- **OPEN SOFTWARE?** No
- **OPEN HARDWARE?** No

Hep Svadja

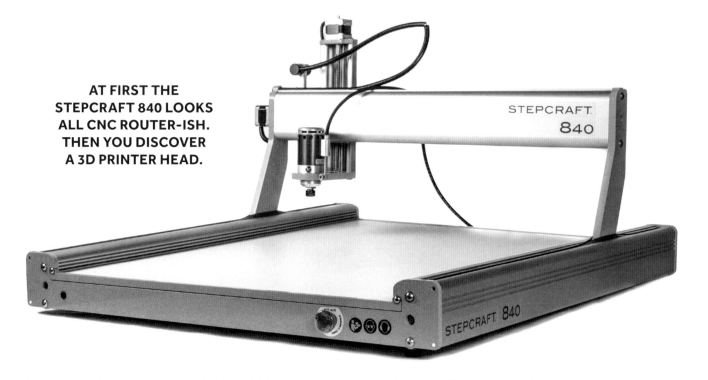

AT FIRST THE STEPCRAFT 840 LOOKS ALL CNC ROUTER-ISH. THEN YOU DISCOVER A 3D PRINTER HEAD.

USCUTTER TITAN 2

This easy to use cutter can tackle a wide range of materials   Written by Mandy L. Stultz

IN OUR TESTING, I FOUND THE USCUTTER TITAN 2'S SOFTWARE A BREEZE, AND ITS SETUP FAST. Using the machine proved to be even easier. This vinyl cutter includes a small ballpoint tip for use when drawing, but the cutting capabilities are where the Titan 2 really shines. It can handle standard vinyl, heat transfer vinyl, cardstock, paint mask, laminate, reflective vinyl, and window film.

Onboard controls make cutting alignment a cinch, and allow for fast maneuvering of material. Test cuts are quick and easy, and the alignment, speed, and force adjustments are at your fingertips.

SMALL ISSUES

I'm not a fan of the plastic tightening screw used to clamp in the blade. It's easy to over-tighten. A quick re-tightening works, but it's hard to tell when is too much.

I found the width ruler/alignment strip easy to measure with, but extremely difficult to use for aligning smaller off-roll pieces of vinyl. This machine could really benefit from longer guide lines to assist with this.

Otherwise, initial material load is relatively painless. I did find the pinch rollers tough to adjust for smaller materials, and no amount of adjustment helped to free them from sticking on the rail. Because the top pinch rollers align with individual bottom rollers instead of a bar or single long roller, a slightly off alignment dramatically skewed any material passing through that was not on a roll.

FLEXIBLE SOFTWARE

The VinylMaster Cut software has far more features than some vinyl cutting software I've used. Letters, shapes, and some clip art images are included. It also allows for super easy importing of a wide variety of image file formats, and provides quite precise fast-tracing capabilities with the Vectorizer tool. The Vectorizer tool also allows for tracing of individual colors inside the image file — fabulous for multilayer/multicolor application — and with the ability to include registration marks in the cut, it makes alignment a snap.

A VERSATILE TOOL

Overall, this is a solid, easy-to-use vinyl cutter for the beginner or pro. While the tested version would be a bit small for professionals, larger versions of the machine could serve busy shops well. ◢

Base price **$995**

PRO TIPS
This machine will cut very fine details with ease. However, sometimes tiny bits of vinyl from those details can become trapped inside the blade housing. To fix, remove and clean the blade and blade housing.

WHY TO BUY
Simple but precise, the USCutter Titan 2 vinyl cutter is very user friendly. Additional perks, like the stand with roll holder and catch basket, keep your material clean and tidy.

- **WEBSITE**
  uscutter.com
- **MANUFACTURER**
  USCutter
- **CUTTING SIZE**
  609×7,620mm, maximum media 711mm wide
- **CUT UNTETHERED?**
  No
- **ONBOARD CONTROLS?**
  Yes, test cut, speed and force adjustments, laser, and feed adjustment
- **CONTROL SOFTWARE**
  VinylMaster Cut OEM (for PC); Sure Cuts a Lot Pro (for Mac)
- **OS**
  Windows XP, Windows Vista, Windows 7 or Windows 8, and Mac OSX
- **OPEN SOFTWARE?**
  No

Hep Svadja

TEST CUTS ARE QUICK AND EASY, AND THE ALIGNMENT, SPEED, AND FORCE ADJUSTMENTS ARE AT YOUR FINGERTIPS

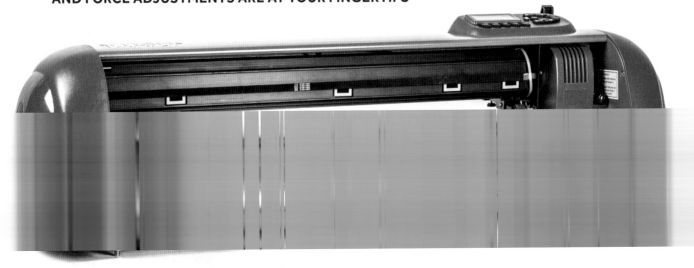

SILHOUETTE CURIO

This versatile cutter is perfect for home use   Written by Mandy L. Stultz

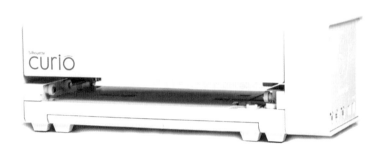

WE'VE LIKED SILHOUETTE'S OTHER MACHINES, BUT THE CURIO UPS THE ANTE WITH NEW TOOLS and the ability to work with thicker materials. The software is user friendly, and Silhouette's website has a nice selection of projects, tips, and tricks.

HELPFUL PROGRAMMING

The Curio comes with a set of platforms that set correct heights for different materials. There are a variety of combinations, but the software tells you how many to use. It can be a bit noisy and rattle as the carriage moves. This didn't seem to affect cutting, but time will tell if it affects performance.

FOR THE HOBBYIST

The Curio may not hold up well in large productions or makerspaces. The blade/pen carriage isn't as rugged as you'd expect for the required tasks, and the cutting mats and snaps that hold them down seem like they'd wear quickly with heavy use. However this is a great machine for the home crafter. ◆

Base price $250

PRO TIPS

Different materials require different combinations of plates and mats to cut and emboss correctly. While the Silhouette Studio software tells you which combination to use when loading up the machine, I found some initial background knowledge made pre-craft setup easier.

WHY TO BUY

The Silhouette Curio does a nice job of cutting a wide variety of materials, from thin vinyl to canvas. It also has the ability to draw, etch, as well as stipple into thin metal sheets, making it a very versatile crafting tool.

- ■ **WEBSITE** silhouetteamerica.com
- ■ **MANUFACTURER** Silhouette
- ■ **CUTTING SIZE** 216×152mm, upgradable to 216×305mm
- ■ **CUT UNTETHERED?** No
- ■ **ONBOARD CONTROLS?** Yes, on/off, pause, and load/unload base plate
- ■ **CONTROL SOFTWARE** Silhouette Studio
- ■ **OS** Windows 7, 8, 10 or Mac OSX 10.7 or higher
- ■ **OPEN SOFTWARE?** No

BROTHER SCANNCUT2 CM350

Use your own designs with this handy machine   Written by Mandy L. Stultz

THE BROTHER SCANNCUT2 WILL CUT PAPER, VINYL, STICKER PAPER, FABRIC, and more. Some accessories are included in the box, but grab a USB Type A to Type B cable or a USB flash drive ahead of time. You can purchase the ScanNCut activation card to connect to Wi-Fi.

CUSTOM DESIGNS

One standout feature is the ability to scan a design (in color or B&W) and then cut it out. Freehand drawings work well, as do preprinted and stamped designs.

Be careful: The stickiness of a brand new mat can ruin thinner materials during removal. Make your first few cuts on sturdier materials if you can. Also, the machine makes an angular "practice cut" along the top of the mat prior to starting. This is likely dulling the blade and could eventually ruin the mat.

QUIRKY BUT CAPABLE

Aside from some small quirks, this machine is solid for any maker, and proved itself easy to learn and utilize for all mediums. ◆

Base price $299

PRO TIPS

Draw your design on your intended paper prior to adhering to the cutting mat. Markers and thick pens will bleed through onto the sticky mat and can create a big mess for future paper.

WHY TO BUY

Scan, cut, draw, and even emboss like a pro with the Brother ScanNCut2. Get started with the library of onboard designs, and tips, tricks, and videos, then shift gears easily into full projects and your own designs.

- ■ **WEBSITE** brother-usa.com
- ■ **MANUFACTURER** Brother
- ■ **CUTTING SIZE** 305×305mm (actual 298×298mm) on provided mat
- ■ **CUT UNTETHERED?** Yes, with purchase of the ScanNCut online activation card
- ■ **ONBOARD CONTROLS?** Yes, touchscreen
- ■ **CONTROL SOFTWARE** ScanNCutCanvas, cloud-based
- ■ **OS** Any with internet access, cloud-based
- ■ **OPEN SOFTWARE?** No

Hep Svadja

ONES TO WATCH

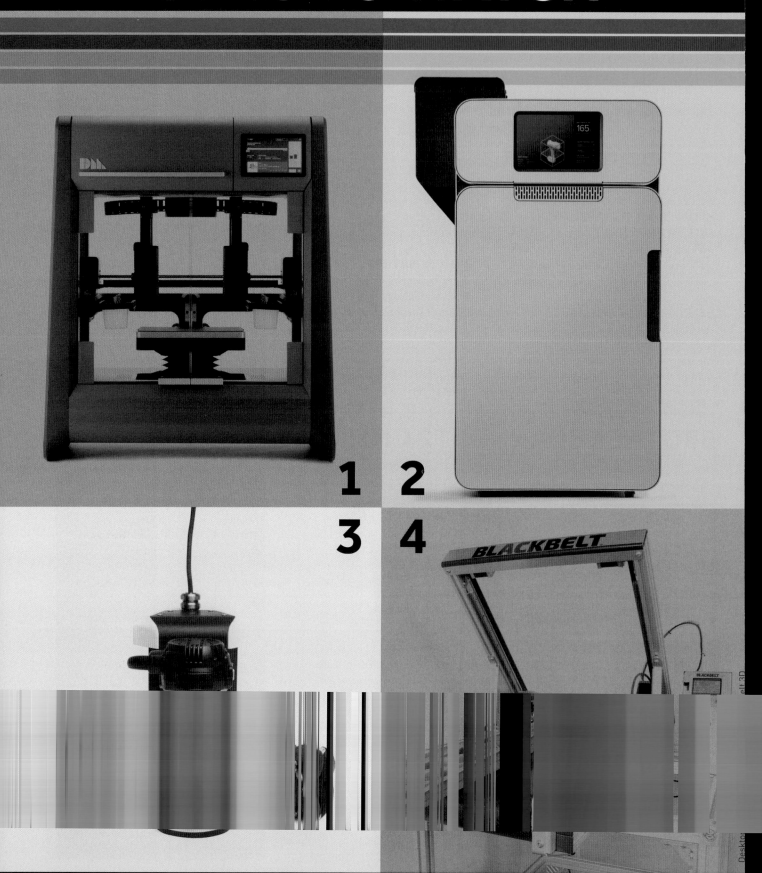

1

2

3

4

BLACKBELT

These innovative new machines show great promise   Written by Matt Stultz

EACH YEAR A SLEW OF ANNOUNCEMENTS COME OUT ABOUT NEW 3D PRINTERS and other DigiFab goodies coming to market. We always try to highlight a few that have excited us and we think you should keep an eye out for. Unfortunately, those machines are not always as quick to come to market as we hope and some of our past "Ones to Watch" are still teasing us with possible delivery.

That said, the latest soon-to-come machines promise exciting aspects for a variety of users. We look forward to testing them all. ◓

1. DESKTOP METAL STUDIO SYSTEM BY DESKTOP METAL

desktopmetal.com/products/studio

Every time I do a talk on 3D printing someone asks "Yeah but can I print in metal?" — often to try to put down 3D printing. If Desktop Metal comes to market, and they look like they will, the answer will soon be YES!

3D printing in metal has been around a long time, mostly done with very expensive powder-based machines. Desktop Metal uses a system far more similar to FDM with metal powders contained in a binder in rod form. Once an object is printed, the binder can be melted away and the part fused in a kiln, leaving you with a metal part ready to be used. This system is not going to be cheap, but in comparison it's a breakthrough. (You may also want to keep your eyes on the Metal X from Markforged while you're at it.)

2. FUSE 1 BY FORMLABS

formlabs.com/3d-printers/fuse-1

Announced earlier this year, the Fuse 1 is Formlabs' first departure from the desktop SLA market that they've come to dominate. The Fuse 1 is an SLS, a printer that — rather than squirting out hot plastic or curing resin with lasers — uses similar lasers to melt and fuse plastic powders. SLS has a distinct advantage over many other 3D printing processes in that any unused material works as a support material for other parts, allowing very complicated shapes to be created. We've covered a few other SLS machines in our Ones to Watch in the past, but with the power of Formlabs behind it, we expect the Fuse 1 will be a big hit next year.

3. GOLIATH CNC BY SPRINGA

goliathcnc.com

We first saw the Goliath at Maker Faire Bay Area this year. The Goliath presents a unique new design for a CNC that doesn't require a large space to setup. The motion of the Goliath is created using omni wheels on the base of the unit, and a router spindle can move up and down in the center to make the cuts. Position control is done by using two guide reels that spool a string in and out, measuring the length from two known points using rotary encoders. The entire package is only about a foot and a half cubed, but can be set up to cut full sheets of plywood.

4. BLACKBELT BY BLACKBELT 3D

blackbelt-3d.com

In 2010, MakerBot announced the Automated Build Platform for their Cupcake CNC 3D printer. This was one of the first systems on the market that made it possible to 3D print objects without user interaction in a production mode. Now, BlackBelt 3D has taken that process one step further, not only allowing parts to be moved on the conveyor belt but replacing one of the axes on the printer with the belt itself. This makes for an almost unlimited build length on the axis that the belt travels. You can add a grab basket to the end to catch small parts that roll off the bed one after another when complete, so you can collect them later. This idea excited the community so much that we have already seen clones being created by companies like Printrbot, trying to truly turn 3D printing into home factories.

BY THE NUMBERS

Compare FFF scores and machines specs in a few simple charts

With such a wide range of options, picking the perfect printer or cutter for your exact needs can be daunting. Let us help you sort through all the data with the following score chart and data tables. The scores below cover every FFF printer in this issue, while the tables on the following pages also include the top machines from the past two years. Find even more online: makezine.com/go/fab-guide-2018.

FFF PRINTER TEST SCORES

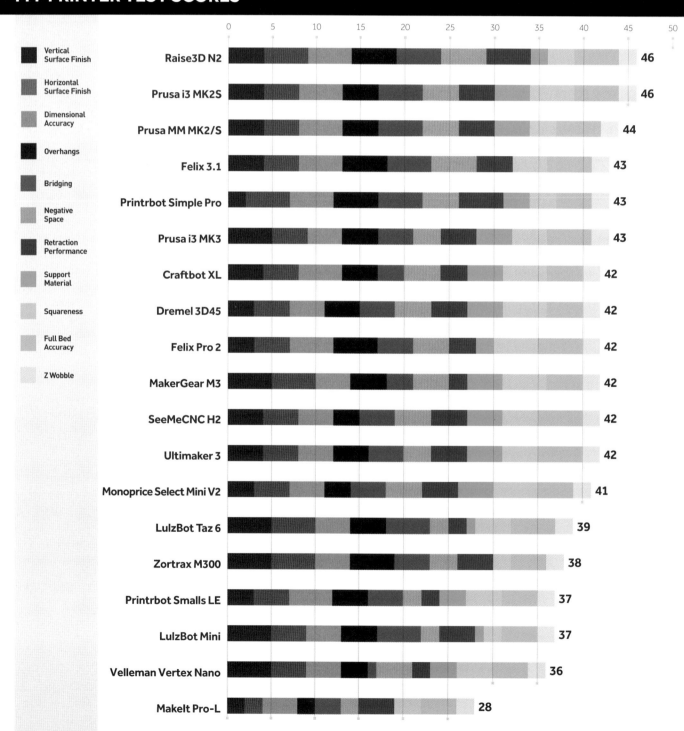

Legend:
- Vertical Surface Finish
- Horizontal Surface Finish
- Dimensional Accuracy
- Overhangs
- Bridging
- Negative Space
- Retraction Performance
- Support Material
- Squareness
- Full Bed Accuracy
- Z Wobble

Printer	Score
Raise3D N2	46
Prusa i3 MK2S	46
Prusa MM MK2/S	44
Felix 3.1	43
Printrbot Simple Pro	43
Prusa i3 MK3	43
Craftbot XL	42
Dremel 3D45	42
Felix Pro 2	42
MakerGear M3	42
SeeMeCNC H2	42
Ultimaker 3	42
Monoprice Select Mini V2	41
LulzBot Taz 6	39
Zortrax M300	38
Printrbot Smalls LE	37
LulzBot Mini	37
Velleman Vertex Nano	36
MakeIt Pro-L	28

FFF COMPARISON

Machine	Manufacturer	Price	Build Volume	Open Filament	Heated Bed	Wi-Fi	Open Source	Review
BCN3D Sigma 2017	BCN3D	$3,132	210×297×210mm		✓		✓	Vol. 58
Craftbot XL	Craft Unique	$1,899	300×200×440mm	✓	✓	✓		page 39
DP200 3DWOX	Sindoh	$1,045	210×200×195mm		✓	✓		Vol. 54
Dremel 3D40	Dremel	$1,599	255×155×170mm			✓		Vol. 54
Dremel 3D45	Dremel	$1,799	255×155×170mm	✓	✓	✓		page 41
Felix 3.1	Felix Printers	$2,150	240×205×225mm (dual extrusion)	✓	✓			page 37
Felix Pro 2	Felix Printers	$2,840	237×244×235mm (dual extrusion)	✓	✓			page 39
Hacker H2	SeeMeCNC	$549	175 (dia.)×200mm OR 140 (dia.)×295mm	✓			✓	page 37
Prusa i3 MK2S	Prusa Research	$599 (kit); $899 (assembled)	250×210×200mm	✓	✓		✓	page 34
Prusa i3 MK2S MM	Prusa Research	$299 (for add-on kit only); $1,198 (as tested)	250×210×200mm (quad extrusion)	✓	✓		✓	page 35
Prusa i3 MK3	Prusa Research	$749 (kit); $999 (assembled)	250×210×200mm	✓	✓	✓	✓	page 35
Jellybox	IMade3D	$799 (kit)	170×160×150mm	✓				Vol. 54
LulzBot Mini	LulzBot	$1,250	152×152×158mm	✓	✓		✓	page 43
M3	MakerGear	$2,550	203×254×203mm	✓	✓	✓	✓	page 40
M300	Zortrax	$3,990	300×300×300mm	✓	✓			page 42
MakeIt Pro-L	MakeIt	$4,399	305×254×330mm (dual extrusion)	✓	✓			page 44
N2	Raise3D	$2,999	305×305×305mm (dual extrusion)	✓	✓	✓		page 32
Printrbot Simple Pro	Printrbot	$699	200×150×200mm	✓	✓	✓	✓	page 36
Printrbot Smalls Limited Edition	Printrbot	$500	173×150×148mm	✓			✓	page 43
Replicator+	MakerBot	$2,499	295×195165mm			✓		Vol. 57
Select Mini V2	Monoprice	$220	120×120×120mm	✓	✓	✓		page 41
Taz 6	LulzBot	$2,500	280×280×250mm	✓	✓		✓	page 42
Ultimaker 2+	Ultimaker	$2,499	223×223×205mm	✓	✓		✓	Vol. 51
Ultimaker 2 Extended+	Ultimaker	$2,999	223×223×304mm	✓	✓		✓	Vol. 54
Ultimaker 2 Go	Ultimaker	$1,199	120×120×115mm	✓	✓		✓	Vol. 48
Ultimaker 3	Ultimaker	$3,495	176×182×200mm (dual extrusion)	✓	✓	✓	✓	page 38
Up Box+	Tiertime	$1,899	255×205×205mm		✓	✓		Vol. 54
Vertex Nano	Velleman	$349 (kit)	80×80×75mm	✓				page 44

SLA COMPARISON

Machine	Manufacturer	Price	Build Volume	Style	Open Resin	Print Untethered	Review
DLP Pro+	mUVe 3D	$1,899	175×98.5×250mm	DLP	✓	✓	Vol. 54
DropLit v2	SeeMeCNC	$740	115×70×115mm	DLP	✓	✓	Vol. 54
Duplicator 7	Wanhao	$495	120×68×200mm	DLP	✓		page 47
Form 2	Formlabs	$3,499	145×145×175mm	SLA	✓	✓	Vol. 48
LittleRP	LittleRP	$599	60×40×100mm	DLP	✓		Vol.48
Moai	Peopoly	$1,250	130×130×180mm	SLA	✓	✓	page 47
Nobel 1.0	XYZprinting	$1,499	128×128×200mm	SLA		✓	Vol. 48
Titan 1	Kudo3D	$3,208	192×108×243mm	DLP	✓		Vol. 48
XFab	DWS Lab	$6,000	180×180mm	SLA			page 46

CNC COMPARISON

Machine	Manufacturer	Base Price	Price as Tested	Build Volume	CAM Software	Materials Handled	Review
Asteroid	Probotix	$3,649	$4,178	635×939.8×127mm	Vectric Cut2D, Vectric PhotoVCarve, MeshCAM, Cut3D, VCarve Pro	Wood/plastics/ soft metals	page 49
Benchtop Pro	CNC Router Parts	$3,250	$4,062	635×635×171mm	Fusion 360	Wood/plastics/ soft metals	page 48
Handibot 2	ShopBot	$3,195	$3,195	152×203×76mm	VCarve Pro ShopBot Edition	Wood/plastics/ soft metals	page 51
High-Z S400T	CNC-Step	$5,299	$14,730	400×300×110mm	Any CAD package should work	Wood/plastics/ soft metals	page 51
Nomad 883 Pro	Carbide 3D	$2,699	$3,100+	203×203×x76mm	Carbide Create or MeshCAM	Wood/plastics/ PCBs/soft metals	Vol. 54
PCNC 440	Tormach	$4,950	$9,895	254×158×254mm	Fusion 360	Wood/plastics/ soft and hard metals	Vol. 54
PRO4824	CNC Router Parts	$3,500	$7,637	1219×609×203mm	VCarve Pro	Wood/plastics/ soft metals	Vol. 54
Shapeoko XXL	Carbide 3D	$1,730	$1,730	838×838×76mm	Carbide Create or MeshCAM	Wood/plastics/ PCBs/soft metals	Vol. 54
ShopBot Desktop Max	ShopBot	$9,090	$9,285	965×635×140mm	VCarve Pro	Wood/plastics/ soft metals	Vol. 54
Sienci Mill One Kit V2	Sienci Labs	$399	$498	235×185×100mm	Universal GCode Sender, any G-code sending software	Wood/plastics/ PCBs/soft metals	page 50
X-Carve	Inventables	$1,329	$1,493	750×750×67mm	Easel	Wood/plastics/ PCBs/soft metals	Vol. 54

LASER COMPARISON

Machine	Manufacturer	Price	Cutting Size	Control Software	Review
Glowforge	Glowforge	$2,995	290×515mm	Glowforge	page 52
Muse	Full Spectrum Laser	$5,000	508×305mm	RetinaEngrave v2	page 52
Voccell DLS	Voccell	$4,999	545×349.25×114mm	Vlaser	Vol. 54

HYBRID COMPARISON

Machine	Manufacturer	Price	Build Volume	Tool Heads	Work Untethered	Open Source	Review
BoXZY	BoXZY	$3,599	165×165×165mm	3D printing extruder, laser module, CNC milling router			Vol. 54
Stepcraft 2/840	Stepcraft	$2,199	600×840×140mm	Milling, 3D printing, engraving, laser etcher, wood burner, hot wire, drag knife, plotting pen			page 53
ZMorph 2.0 SX	ZMorph	$3,890	250×235×165mm (with covers) 300×235×165mm (open)	3D printing, dual extruder, paste extruder, milling head, laser module	✓		Vol. 54

VINYL CUTTER COMPARISON

Machine	Manufacturer	Price	Cutting Size	Cut Untethered	Control Software	Review
CAMM-1 GS-24	Roland	$1,995	584×25,000mm		Roland OnSupport; Roland CutStudio	Vol. 54
Curio	Silhouette	$250	216×152mm		Silhouette Studio	page 55
MH871-MK2	USCutter	$290	780mm×roll length		Sure Cuts A Lot Pro	Vol. 48
Portrait	Silhouette	$199	203×305mm		Silhouette Studio	Vol. 48
ScanNCut2 CM350	Brother	$299	298×298mm	✓	ScanNCutCanvas	page 55
Cameo	Silhouette	$299	305×305mm	✓	Silhouette Studio	Vol. 48
Titan 2	USCutter	$995	609×7,620mm		VinylMaster Cut OEM (PC); Sure Cuts a Lot Pro (Mac)	page 54

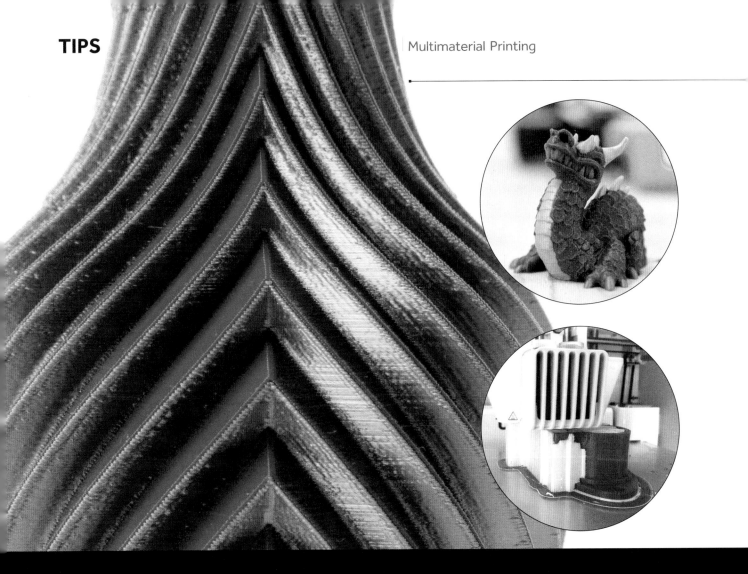

THE FUTURE IS BRIGHT

Combining multiple materials brings a new dimension to 3D printing Written by Matt Griffin

While producing parts with more than one material might not be a new concept for industrial additive manufacturing, it's been an area of development for desktop 3D printing for many years. Good news: manufacturers are now offering machines with multiple toolheads, toolheads to mix multiple filament sources, and secondary marking/staining subsystems.

Why might you want to consider multimaterial printing? Here are a few key areas that offer exciting new capabilities.

PRINTABLE SUPPORT MATERIALS

Tired of changing your design to avoid overhangs, fragile surface features, and close-fitting parts that fuse across intentional gaps? A secondary material that you can dissolve, pull away, or break off from the intended part not only saves countless hours of post-processing, but also permits new and previously unproducible designs. Thanks to these sacrificial materials, 3DP makers can produce parts with interior chambers, precision print-in-place articulations, and tiny text markers, expanding the capability of an affordable printer. Polyvinyl alcohol (PVA, i.e. "wood glue" and similar variants) can be dissolved in tap water, HIPS can be dissolved in limonene (be careful!), and special thermoplastic polyurethane (TPU) materials can be peeled away by hand.

HARD, SQUISHY, OR HARD-AND-SQUISHY?

3D printing technology relies on the special-case, non-Newtonian, extrusion-friendly properties associated with thermoplastic materials, such as "shear thinning." While these special plastics can't do everything, you can accomplish a lot by picking variants with differences in durometer ("rigidity"), especially when adding in thermoplastics' near cousins, thermoplastic elastomers (TPEs). One can achieve functional parts by combining rigid and rubbery elements into the same parts. Living hinges, grip-friendly handles, load-bearing soft robotics, and interactive interfaces are a few of the most popular projects explored by 3DP makers, thanks to the availability of timing belt–grade TPUs such as NinjaFlex and easier-to-extrude semiflex materials such as harder TPUs, nylons, polyesters, and rubberized PLAs.

WELDING LIKE TO LIKE

Expert operators are starting to recognize how the *same* material on multiple heads/feeds can also open your toolbox to new strategies. Thanks to slicer tools such as Cura, Simplify3D, and variants of Slic3r, exporting a single part made up of multiple settings-assignable bodies has never been easier. Operators are able to assign just what they need to each element of their design by installing different nozzles/cores and tweaking temperature and speed settings. It's a vast improvement over the compromise of selecting a single setting for an entire object. As slicers offer increasingly targeted, feature-driven strategies (such as variable infill, feature recognition, and nuanced flow control), and as composite materials offer a range of properties that can be expressed through small fluctuations in mass or temperature, this category of "multihead" (vs. "multimaterial") printing may prove instrumental in advancing FFF technology.

EMBEDDED CONDUCTIVE MATERIALS

Multimaterial printing strategies offer opportunities to make conductive materials useful and interesting despite their inherent high resistance values. Embedding conductive materials into the surface of your parts can allow a sensor placement in an area difficult to populate with off-the-shelf electronic components. And while the goal for low-resistance, high-current power routing still tends to require specialized extrusion and processing hardware, the opportunities to exploit the electrostatic discharge safety features of low-conductive materials have immediate value to those using desktop 3D printers to produce electronics enclosures, manufacturing aids, and testers.

A PANOPLY OF COLORS

The color element remains a creative opportunity for expression and communication. When cleverly planned, even increasing the palette of your plastic paint brush to two heads can transform your design. There have always been desktop filament providers aiming to offer 3DP makers a big crayon box to work with, but new offerings — a vast range of skin-matching colors from Eastman, matte and sparkly Proto-pasta filaments, and a metallic shimmer with Polyalchemy's Elixir PLAs — are great on their own, and powerful in combination. ◐

MATT GRIFFIN is the director of community for Ultimaker North America. He is also a writer, teacher, and consultant covering topics such as 3D printing, electronics hobbyists, and more.

TIPS

» Never done this before? Recreate a known example design to uncover any tweaks needed for your software/hardware/material combination instead of learning with your brand-new object.

» Are your materials compatible? Check data sheets first, or you might have bonding difficulties later.

» Create a tiny test swatch that you can print quickly to tune your process without wasting a lot of plastic and time. (I print little robots!)

» Use 3MF files or similar fabrication-friendly formats to export more than one assignable part within a single container file. This is great for archiving as well.

» If you need to export separate STLs for each element, make sure to name them clearly to help you load and assign them correctly.

» Cura, Simplify3D, and versions of Slic3r can all handle multiple materials — check to see which route is best for you and your hardware.

» Make sure your 3D design software exports separate geometry for each color, not just surface color.

» Software like Cura can recall the placement assigned by your design software, making it possible to lock multiple parts into position automatically without needing to make adjustments in the 3D control software.

MULTIMATERIAL PRINTING WITH A SINGLE EXTRUDER!

Mosaic's Palette+ enables multimaterial printing on nearly all 1.75mm filament-based printers — no modifications required. The device measures and precisely splices various filament that feed live into your printer to create colorful, flexible, soluble, and rigid materials, all in a single part.

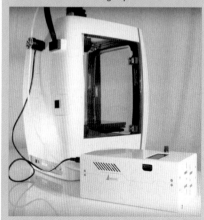

PICK YOUR NOZZLE

Custom hot ends let you print with specialty filaments without damaging your hardware or degrading print performance Written by Matt Stultz

Over the past year we have been running regular Filament Friday articles on makezine.com, showing you some of the growing options for materials you can feed into your printer. We've covered shiny filaments, flexible filaments, clear filaments, and recycled filaments.

One category of filament we have been avoiding though have been "fill" filaments; PLA, ABS, or nylon blends with other ingredients like metal powders or carbon fiber included in them. The reason for this omission is because on a stock 3D printer, they can damage your nozzle and degrade your print performance.

Thankfully there is a solution: custom nozzles that are harder than the standard brass and resist wear from abrasive materials. Depending on your machine and your budget there are a range of nozzles available for you. The good news is many manufacturers use the same thread fitting so there is a good chance your machine is compatible. We got all the nozzles we tested from matterhackers.com.

TYPES OF TIPS

BRASS

Brass is your standard nozzle material. It has great thermal characteristics and is easy to machine because it's soft — which means it is very easy to grind away with materials like carbon fiber filled nylon.

STAINLESS STEEL

Stainless is often used for precision applications like medical equipment, machinery, and cutlery because it is hard and resists rusting or other corrosion. It is much harder to machine than brass so parts made from it come at a higher cost but this is a great step towards protecting your nozzle from wear.

HARDENED STEEL

To make hardened steel nozzles, a stainless nozzle is heat-treated and coated to make it even less likely to wear with highly abrasive materials. If you are going to be printing in carbon fiber filaments a lot, this is a must and is my top pick for the nozzle to upgrade to.

RUBY

Rubies and sapphires come in just behind diamonds on the Mohs scale of mineral hardness. They can be polished smooth and have great thermal properties. They have been used for years as the pivot points in expensive watches because they don't wear and have low friction. All of this makes them perfect for use as 3D printing nozzles ... if you have the budget. ⊘

Hep Svadja, MatterHackers, Proto-pasta

TWEAK,
HACK,
& BEND...

6 ISSUES FOR ONLY $34.99!

SAVE 42% OFF THE COVER PRICE

INCLUDES
DIGITAL
EDITION

B78NS7

NAME _____
(please print)

ADDRESS/APT. _____

CITY/STATE/ ZIP _____

COUNTRY _____

EMAIL ADDRESS (required for order confirmation and access to digital edition)
_____

☐ Payment Enclosed   ☐ Bill Me Later

FOR FASTER SERVICE, GO TO:
MAKEZINE.COM/ORDER

Make: is published bimonthly. Allow 4-6 weeks for delivery of your first issue. For Canada, add $9 per subscription. For orders outside of the U.S. and Canada, add $15. Payable in U.S. funds only. Access your digital edition after receipt of payment at make-digital.com.

FANCY FILAMENTS

So you are going to bite the bullet and upgrade your extruder — what should you print with your new hardened hot end?

NYLON X

makezine.com/go/nylon-x-filament

Nylon is a great material for 3D printing. It holds up to wear and tear, it's strong and takes a lot more to break, but it tends to be a bit too flexible for lots of applications. Nylon X helps solve that problem. Take all the great qualities of nylon and stiffen it up by the inclusion of carbon fiber.

MAGNETIC IRON

makezine.com/go/magnetic-iron-filament

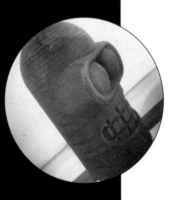

We've mentioned the magnetic iron PLA from Proto-pasta before but it's worth mentioning again because it's great. The ability to rust your prints after printing using various process can really step up the aesthetics.

GLOWFILL

makezine.com/go/glowfill-filament

Everyone loves things that glow in the dark. ColorFabb's Glowfill PLA/PHA blend is one of the best glowing filaments out there. Like many of them though, the material added to make it glow (strontium aluminate) is abrasive so a new nozzle is useful to keep your machine safe.

MATT STULTZ
is the 3D printing and digital fabrication lead for *Make:*. He is also the founder and organizer of 3DPPVD and Ocean State Maker Mill, where he spends his time tinkering in Rhode Island.

THE MANE ATTRACTION

Ingenuity in the printing process has led to an explosion of models with hair

Written by Caleb Kraft

IN some ways, 3D printer machine design has slowed a bit. Most of the new releases are iterative improvements on the previous models — a normal occurrence in a maturing market. If you're seeking a bit of "wow" though, there is one place you can look: new printing methods.

BRIDGING THE GAP

As printers are getting more capable and reliable, the community has been coming up with interesting methods of using the machines for peculiar results. "Bridging," or stringing molten filament across a considerable gap without support below it, was once something that only the best-tuned printers could pull off. Now, pretty much any printer can bridge a decent gap with some degree of reliability.

Over the past couple years, people have taken the concept of bridging and twisted it into a way of creating hairy or furry models. There's an entire collection on Thingiverse of hairy models that you can simply download and print.

Some of these examples go back a few years, and are often fairly rudimentary, like the "drooloop" flowers by Mark Peeters (thingiverse.com/thing:240158), shown here in the bottom left photo.

Over the last year or so, there has been an explosion of designs that really utilize this process to incredible effect. Two of the most notable are the Hairy Lion (thingiverse.com/thing:2007221), which has a big flowing mane, and Hairy Einstein (thingiverse.com/thing:2151104, shown here in bottom right photo), complete with a disheveled 'do.

These aren't simply files that you can print and be done with, needing a bit of post-processing attention. They print with an additional structure that serves to be the target of the "bridge." When they're done printing, you snap away that extra structure and then "style" the hair with a heat gun. This allows for the fantastic look to be custom and unique to each print.

After seeing what incredible things people are doing with something as mundane as bridging a gap, I'm curious what will be the next area for ingenuity. ◉

CALEB KRAFT is senior editor for *Make:* and has had a 3D printer since 2012. He is delighted to see printable designs that incorporate a tiny bit of chaos and disorder.

Hep Svadja, Mark Peeters, Jwall

FAITHFUL FUSER

Weld parts and more with a 3D printing pen

Written by Chris Yohe

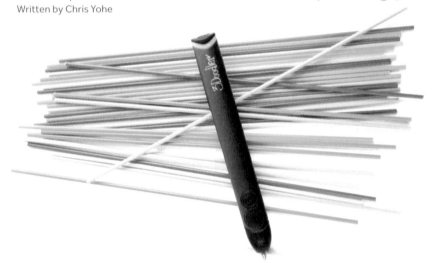

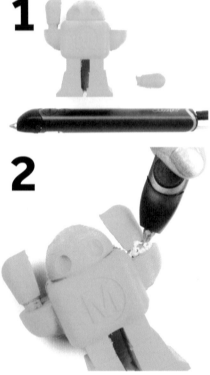

1

2

3

If you spend any time in 3D printing, you will likely succumb to the lure of the 3D printing pen. Basically an extruder mounted in an oversized marker, they have gone from a little pricey to the point where you can find them for less than the cost of a movie ticket, removing a lot of the barrier to entry. At the impulse-buy price the only thing stopping me was the thought of how hard it is to get great prints with a finely tuned machine, let alone out of thin air. But once you realize the practical applications, one of these gadgets will surely become a go-to tool for you too.

JOINING TWO PIECES

One of the best and easiest uses of the pen is to "weld" two 3D printed parts (**Figure 1**). First power on your pen and warm it up to temperature. Most are PLA-only but some now do have other options, depending on the model. Select a piece of PLA sized appropriately for your device, and (typically) in the color of the parts you are trying to join. Hold or clamp your pieces together, making sure to have good access to your seam. Placing the tip firmly against the seam, extrude the filament into it (**Figure 2**), making multiple passes if needed. It will appear much like a horrible-looking first layer, but have no fear. Once you have finished welding the whole seam, take a pair of wire cutters or scissors and trim off any large blobs and beads.

Carefully go over it with a hobby knife, scalpel, or even a small plane. Take your time to get it as smooth as possible, without damaging the print itself. After the rough trim, you can then use sandpaper in increasing grits to lightly sand down the seam to fully smooth (**Figure 3**). If you discover any gaps or rough areas, you can fill them in by repeating these methods. Surface whitening from sanding can often be helped with a little heat from a heat gun or other source.

OTHER USES

This method is one of the best since you are joining two parts with the exact same material they are made of, making them one strong, solid piece. It also has the added bonus of helping you justify that impulse buy to those who mocked you!

But there are many other real-world uses — that same hot end can be applied to melt and remove or smooth blobs on the surface or rough support stubble, as well as discolorations. Another hot recent trend has been micro sanding, so you can give a low-tech version of that a shot too. ◙

CHRIS YOHE
is a professional software developer and cheap digifab freak. He co-founded 3DPPGH and is a member at HackPittsburgh.

Hep Svadja

CALCULATED CUTTING

Written by Clement Moreau

5 key elements to keep in mind for your laser cut design

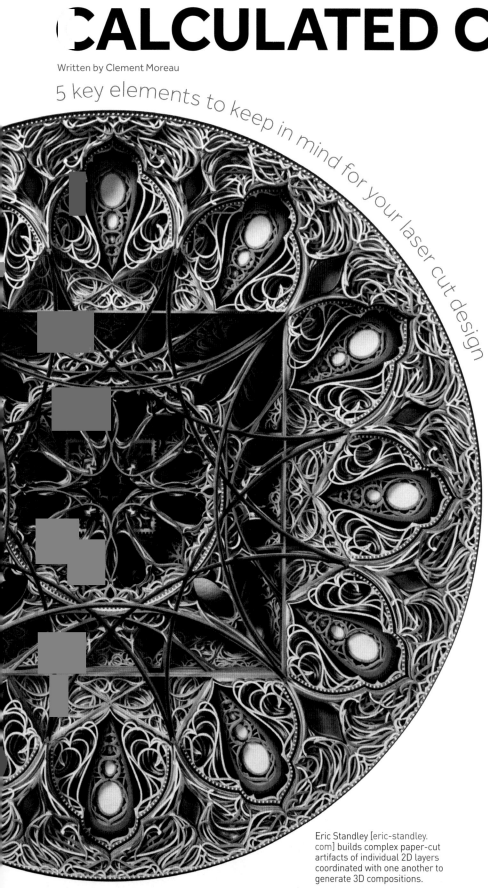

Eric Standley [eric-standley.
com] builds complex paper-cut
artifacts of individual 2D layers
coordinated with one another to
generate 3D compositions.

We've recently added laser cutting (sculpteo.com/en/lasercutting) to our offerings at Sculpteo. While it offers many advantages such as precision, strong repeatability, and cost-effectiveness, there are a few things you should ponder before trying it for your project. These considerations are important whether you are using our service, or using a home laser cutter.

1. LASER CUTTERS ONLY SUPPORT VECTOR FILES

Simple pictures or non-vectorized drawings are not enough to create an object with laser cutting because they do not contain the information necessary to enable the machines to cut or engrave it. What you need is a vector file.

New to CAD design? You can find ready-made vector files on dedicated websites or ask a designer to do it for you. But it is also very easy to create your own vector file with software such as Adobe Illustrator, Inkscape, or SketchUp. Whatever software you use, you should not forget the design guidelines that must be respected, depending on the material you want to use. Finally, always make sure that your design is in the right format so you can send it to the laser cutting service. For example, at Sculpteo we accept vector files in .SVG, .DXF, .AI or .EPS format.

2. FIND THE RIGHT MATERIAL FOR YOUR PROJECT

Each project is unique and each design needs to find its adequate material to turn into an object. But before diving in, you need to be aware of the different properties of each material. If your object is going to be exposed to humidity or heat, you might want to avoid cardboard or MDF and have a closer look at acrylics. On the contrary, if you have a tight budget or want to try a prototype design, then cardboard can be a good option.

You must also take into account the cut, or kerf, that the laser creates as it passes over the material and how your material will react to the heat of the laser (burn marks, cracks, etc.).

3. TURN A LASER-CUT-FRIENDLY DESIGN INTO A LASER-CUT-PROOF ONE

Once your design is ready, rethink it! Is it suited for laser cutting? Narrow parts, superfluous details, and wide engraved surfaces could damage the final result of your piece or increase the price of your design. Keep in mind that if you optimize the line work, you will achieve faster cuts and save money.

Clever designers will figure out the best way to maximize the use of materials. Look carefully at your design and get rid of anything that could be considered superfluous. You should remove double lines as well. A laser, as powerful and precise as it may be, doesn't have a brain and it doesn't know you don't want to cut the same line twice. It may sound obvious, but be careful when you create your design in software. Fonts must also be vectorized, otherwise they won't be included in the pattern. Finally, we strongly advise reducing raster engraving as much as possible since it is the most time consuming operation.

4. KEEP IN MIND THE KERF WHEN YOU DESIGN FOR ASSEMBLY

One of the good things with laser cutting is that parts that are cut can easily be assembled. In order to do that you must plan for a minimum space between each of the parts. Make sure to build your design in a way that allows space for the kerf.

If you want your objects to fit into each other you must take the kerf into account, which means subtract half the size of the kerf from the frame's parameter, and add the other half of the kerf to the inner part. It takes some time, but that's the only way to prepare a project like Figure 4.

If you wish to make pieces fit into each other and make sure they'll stay connected, you absolutely need to add in nodes. Nodes are small bumps situated in a piece's slots or tabs that allow you to compensate for the thickness variations of the material and the kerf. Nodes get compressed when the pieces are assembled and they concentrate

the friction on specific points rather than on the slot's whole surface. This way, the slots can be larger without coming apart, allowing the pieces to stay together.

In order to make sure the pieces stay fixed, you must place the nodes on each side of the slot opposite to one another. Depending on the length of the slot, you can place several nodes. This will minimize the tension that could potentially occur if the nodes were not aligned or if one was missing. They must be smooth and long enough to ease the interlocking. Depending on the material's density, the nodes' width may be increased or decreased. The higher the material's density, the smaller the nodes' width should be.

5. LASER CUTTING CAN BE GREAT FOR 3D OBJECTS TOO

Laser cutting and 3D printing are both wonderful digital manufacturing techniques. Freedom of design, super low set-up fee, speed ... so many advantages that you can play with. Both techniques represent a powerful tool, useful for designers who now need to rapidly design, prototype, iterate, and produce. And it's a real pleasure to create products that combine these two techniques. But you need to think outside of the box!

As you know, laser cutting allows you to make 2D shapes, but it can also be used as an efficient solution to create 3D objects. Sometimes it can even suit your design idea better than 3D printing would. The use of laser-cut parts for production requires thinking in 2D in order to assemble in 3D. Designing in 3D by using laser cutting may seem a brainteaser, but it can save you a lot of money. ◢

If you want to learn more about laser cutting as well as the different materials and their properties, Sculpteo offers its e-book *The Ultimate Guide to Laser Cutting*, available for free on their website (registration required).

CLEMENT MOREAU
is co-founder and CEO of Sculpteo an online 3D printing and laser-cutting service based in San Francisco and Paris. He holds a MSc. in Engineering from Ecole Centrale de Paris.

If you want to learn how to make the shape featured in this picture, while saving material and respecting the kerf, Sculpteo has specific tutorials on the subject, sculpteo.com/en/lasercutting/prepare-your-file-laser-cutting.

Eric Standley, Hep Svadja, Sculpteo

CUT TO THE CHASE

Why you should buy a laser cutter before getting a 3D printer Written by Caleb Kraft

Laser cutters can cut, etch, or engrave a wide variety of materials, such as fabric, plastic, and glass.

CALEB KRAFT is senior editor for *Make:* and a habitual maker. His cat-like short attention span leads him to seek out the fastest methods of prototyping, and to chase lasers.

For the past 10 years or so that I've been actively writing about the maker movement, I've been lucky enough to witness the birth of many hackerspaces, makerspaces, fab labs, and other communities. During this time, I've seen fads and trends come and go, both in organizational structure and in equipment.

At least a few times each month, someone who is adding a makerspace to their school, library, or local community asks me what fancy piece of equipment I would recommend they start with. My answer often surprises them.

Nearly everyone expects me to say, "Get a 3D printer." Printers are, it seems, omnipresent in any maker related activity these days, and without a doubt, incredible. They can produce things that are impossible by other means. Not only that, they draw a crowd, which can be an important feat on its own when trying to acquire new membership.

However, I do not recommend a 3D printer as a first equipment purchase. If you could only buy one thing, I recommend a laser cutter. There are two main reasons that a laser cutter tops my equipment list:

THEY'RE FAST

A student or maker can design something, send it to the laser, and have it cut in minutes. Even complex engraving is done in a period of time that most people are willing to sit and wait for. The instant gratification is not the only factor at play here; this allows more people to use the machine in a shorter amount of time.

I'm always sad to hear about schools that get a 3D printer, then find that only a single student can do a print per day, assuming that the print doesn't fail and need to be started over again.

THEY'RE SIMPLE

Designing for a laser cutter can start out at a much easier level than with full 3D modeling. Simple doodles can be converted and cut with ease using open source and free tools like Inkscape. Absolute beginners can walk away with professional results in the time it takes to drink a coffee.

You may argue that someone could design fast and simple things for a 3D printer, but at least in my experience, this just doesn't happen. There is a considerable learning curve, and print times almost always come as a shock to beginners. ◔

Hep Svadja/Shot on location with Josh Jakus and Elizabeth Woll of Automatic Arts [automatic-arts.com]

ENGRAVER ENHANCEMENTS
Pump up your cheap K40 laser with these tips and add-ons
Written by Tyler Winegarner

Semi-professional laser cutters have come down in price considerably in the past few years, but $5,000+ is still a pretty significant investment. Then there's the K40 — a 40-watt CO2 laser engraver, usually sold on eBay, with a smallish work area of 8"×12". Costing less than five hundred dollars, many have wondered, "How bad can it be?" and "How good can it get?"

The K40's money savings comes from cheap parts and corner-cutting everywhere, and it offers very little in the way of safety equipment — a fire extinguisher is an absolute necessity when working with any laser. You'll also need a pair of laser safety glasses rated for CO2 lasers. The fume extraction fan doesn't move a large volume of air, and the vent tube is made of thin plastic — both could do with an upgrade, especially if you're going to be cutting a lot of flammable materials.

The K40's software, LaserDRW, presents a steep learning curve regardless of familiarity with other laser cutter software packages. Lining up the modules for cutting and engraving on the same workpiece is a challenge. The software only imports rastered bitmaps, no vector files. Every function has a confusing and non-intuitive label. The laser itself is not controlled through software, rather by a knob on the control panel — something to consider if you're looking for repeatability. Good news, you can upgrade this machine. Some options:

1. REPLACEMENT BED
The sliding clamp system is a simple and effective method of workholding, but it's also small and limits you to only thin, rigid materials. Swap it out for an expanded steel grate and an adjustable z-height system.

2. EXPANDED GANTRY
The housing of the K40 is pretty massive for the 8"×12" work area. But you can expand the gantry and cutting area by relocating the power supply and control board out of the enclosure.

3. K40 WHISPERER
This open source software replaces LaserDRW as the software driver for the stock controller board. It is laid out much more intuitively, allows the importing of vector files, and will work without the hardware dongle. It's still pretty buggy, but is sure to improve over time.

4. SMOOTHIEBOARD
You can also replace the entire controller board to get around software woes. The Smoothieboard offers just that, opening a variety of open-source software options.

Ultimately, you're not going to make this into a Glowforge. But for under $500, you'll have a functioning, upgradable laser cutter, and that's remarkable. If you have time to tinker and don't have commercial production expectations, the K40 could be a nice shop addition. ●

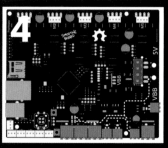

TYLER WINEGARNER is the video producer for *Make:*, as well as a maker, tool user, story teller and skill hoarder. He is driven by the weird and wonderful.

Hep Svadja

Paste extruders can be screw- or pneumatic-based syringes or rotating augers.

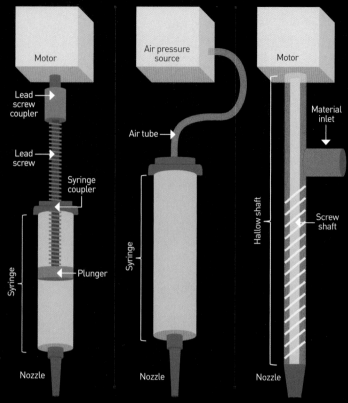

Motor

Lead screw coupler

Lead screw

Syringe coupler

Syringe

Plunger

Nozzle

Air pressure source

Air tube

Syringe

Nozzle

Motor

Material inlet

Hallow shaft

Screw shaft

Nozzle

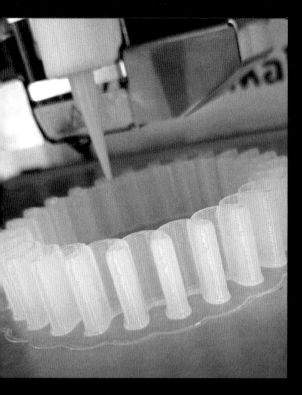

●●PY IN PASTE

Written by Charles Mire

Liquid-based printing offers many options for materials and applications

Plastic materials have largely dominated the overall 3D printing market, simply because they are well understood, are fairly easy to process, and have had immediate practical applications. However, I believe the rise of paste materials is coming.

ADVANTAGES

First, the starting format of paste material is a viscous liquid, rather than a rigid material that must be fabricated into filament. This one key property means that paste materials can have many more variations to their formulations than typical filament. The possible selection of paste materials includes elastomers (silicone, latex, polyurethane), clays (including metal clay, which can be fired post printing to result in a pure metal part), biomaterials, wax, electronic inks, foods, and plastics. Paste materials are often stored in a sealed container, so air exposure is greatly reduced between prints. This translates to a much longer shelf life than filaments.

The applications for paste materials are quite extensive. Rubber is used for shoes and gaskets, silicone for medical devices, and ceramics for special high-temperature

applications. Conductive traces on circuit boards, clothing, or other surfaces can be applied in paste form. Biomaterials are being heavily researched for medical and pharmaceutical applications. This huge area is ripe for advancements in 3D printing.

WORKING WITH PASTE

Some designs work very well for paste printing, while others are challenging. A shoe insole is ideal because it is flat, with minimal tricky parts. Objects like vases start to add some small challenges, especially if they're tall or have bulging portions. Features that don't tend to work well include overhangs, thin walls that must support a lot of weight for tall prints, and lots of irregular shapes combined into one print. These kinds of challenges don't necessarily mean a complicated object can't be printed. One approach to printing

CHARLES MIRE has spent the past 10 years working on 3D printing soft materials, and, aside from running Structur3D Printing, likes to spend time with his family and in the Canadian outdoors.

something tricky is to print sections of the final object and then adhere them together post printing.

Paste materials require slower print speeds than filament. The nozzle size, layer thickness, wall thickness, and infill amount and pattern are all settings you can optimize to achieve high-quality paste prints. We've published some baseline Slic3r settings on our forum (forum.structur3d.io) for some of the main materials we have printed. By keeping good notes, you can develop your own baseline optimizations too.

The material should have enough stiffness to hold a shape. Honey prints poorly, whereas cake icing prints well. We've shared our own variation of the recipe for royal icing on our forum. Food materials can be a great way to get started with paste printing, simply because they are inexpensive and widely available.

Paste printing is similar to using filament, with the benefit of being able to vary your paste material a thousand different ways. The advantages for customization using 3D printing are clear, and it is only a matter of time before paste printing becomes more

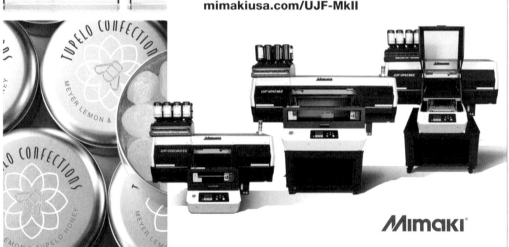

PIMP MY PRINTER

Written by Ryan Priore

Take your dirt-cheap device from ho-hum to hot rod

Although in many instances you get what you pay for, there are a few immediate budget modifications that will transform an entry-level bot into a respectable hot rod. Before you get started on any of these, find the relevant online forum or social media outlet for your printer. Some involve getting your hands dirty, so roll up those sleeves:

1. SAFETY FIRST

Exposed electronics or mains wiring should be placed inside of an appropriate enclosure with adequate air flow. If the only way to power off your bot is by directly unplugging the system, a fused switch should be added (or simply use the on/off function of a power strip). Incorrect/underrated connectors on the controller board (e.g. 10A terminals where a heated bed may draw 11A) should be replaced with appropriately rated connectors.

2. CONSIDER YOUR SOFTWARE

Vendor-recommended slicing software is a suggestion — I find myself often coming back to good ol' Cura. Many community users share their favorite slicing profiles for their specific machines.

3. UPDATE FIRMWARE

When you receive your budget bot, the manufacturer or community may have posted key firmware updates, which can affect printer functions or offer performance tweaks (e.g. acceleration, PID tuning, etc.).

4. COOL YOUR PARTS

PLA loves to be cooled as it prints. Improve your results with a part-cooling fan — plug it into an available software-controllable port on the controller board or add an inline switch to manually turn it on/off.

5. ADD AN LCD INTERFACE WITH SD CARD SUPPORT

Printing while tethered to your desktop or laptop computer is so 2012. Instead, add an LCD interface with an SD card reader for controlling basic printer operations and loading files to print. Popular options for the

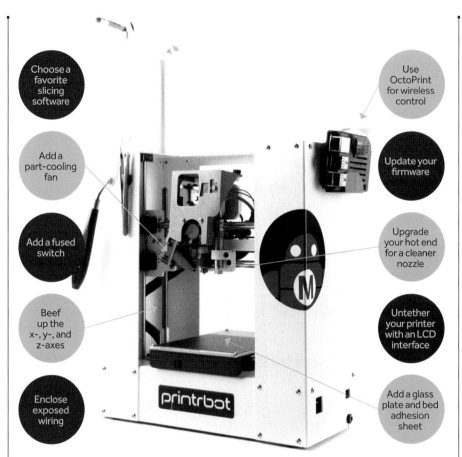

Choose a favorite slicing software

Use OctoPrint for wireless control

Add a part-cooling fan

Update your firmware

Add a fused switch

Upgrade your hot end for a cleaner nozzle

Beef up the x-, y-, and z-axes

Untether your printer with an LCD interface

Enclose exposed wiring

Add a glass plate and bed adhesion sheet

printrbot

RAMPS controller board include the Smart Controller and Full Graphic screens from RepRapDiscount.

6. UPGRADE YOUR HOT END

A legitimate hot end (e.g. E3D v6) can minimize future nozzle clogs. You may need to modify the thermistor settings in your printer firmware and PID-tune your hot end after installation.

7. IMPROVE YOUR PRINT SURFACE

Ensure a flat print surface by installing a glass plate to the printer bed (custom cut from your local hardware store) along with a bed adhesion assistant (e.g. BuildTak, PEI sheet, etc.). You may need to add a spacer to the z-axis stop to compensate for the additional thickness of the new print surface.

8. INSTALL SMOOTHER THREADED RODS

On Cartesian-style machines (which most 3D printers are), swap out those all-thread M5 or M8 rods with Acme rods to clean up any observed banding along the z-axis. Beef up the bearings on the x- and y-axes for buttery-smooth linear movements.

9. GO WIRELESS

Add remote printer control and monitoring by using OctoPrint and a webcam on your favorite single-board computer. ⊘

RYAN PRIORE is a spectroscopist and photonics entrepreneur. He cofounded 3DPPGH and is a member of HackPGH. Ryan jokingly describes himself as a capitalist by day and an open-source enthusiast by night.

Hep Svadja

UP YOUR GAME

Written by Matt Stultz

Give your old machine new life with one of these upgrades

X-Carve dust collector accessory from Inventables

Each year we give you a new list of 3D printers, CNC machines, and other great devices. Sometimes though, the best tool for the job is the one you already have. Here are some updates and upgrades to several machines we have reviewed in the past and still love.

1.75MM CONVERSION FOR THE ULTIMAKER 2 FAMILY

thegr5store.com

The days of 2.85mm filament are numbered and I am on a one-man campaign to end them. There are two manufacturers largely responsible for keeping it alive, Ultimaker and LulzBot. I'm a fan of both's products, and you can successfully run 1.75mm in LulzBot and Ultimaker 2 as-is, but for greater reliability and quality, matching hardware is preferred. Now there is a kit to convert your Ultimaker 2+ family fully to 1.75 filament and be done with 2.85 in one of our favorite brands of printers.

CARBIDE COPPER FROM CARBIDE 3D

copper.carbide3d.com

We've tested the Nomad 883 in years past and love it for a desktop CNC that keeps the mess out of your work area and looks good doing it. This year, Carbide 3D announced a new software package, Carbide Copper. This new app has one purpose: milling PCBs. I gave it a try and found it to be the easiest solution I have ever tried for custom

PCBs on desktop mills. I do wish it had a custom board shape layer that you could add to process the entire job in one piece of software, but since they are offering something that instantly upgrades all of their customers for free, you can't fault them too much.

PRUSA I3 2.5 UPGRADE FROM PRUSA RESEARCH

shop.prusa3d.com

With the announcement of the i3 MK3 from Prusa Research, lots of users are looking at their current MK2/S machines and feeling like they might be missing out. Thankfully the Prusa team has an upgrade kit to help get you some of the new features. The 2.5 kit includes the new bed with removable spring steel sheet, the new filament sensor, the PINDA 2, and a few other goodies to help make your MK2/S printers feel like new again.

NEW EASEL PRO AND DUST COLLECTION FROM INVENTABLES

inventables.com/technologies

One big downside to CNC milling over 3D printing is the amount of mess it makes with dust flying everywhere. The key to a clean shop is a good dust collection system, something that was missing from the X-Carve family of machines out of the box. Now Inventables has created a dust shoe and hose management system that makes it easy to add a vacuum to your X-Carve

and keep it all clean. This is especially important to pro users. Inventables has another update for that group as well, Easel Pro. Their new software package, targeted toward users who are making a living with their X-Carves, contains optimizations and features, like part nesting, for getting the most out of your CNC.

AUTOMATIC TOOL CHANGER FOR THE TORMACH 440 BY TORMACH

makezine.com/go/tormach-tool-changer

Switching tools during a CNC job is a perfect opportunity to make a mistake, and it's time consuming during complex jobs. Tormach has had automated tool changers in the past for their larger models but the garage-friendly 440 was missing this option until now. If you read our review of the 440 in *Make:* Vol. 54, you will know this is a beast in a small(ish) package — the addition of the tool changer really opens doors for those trying to get some serious work done with it. ●

MATT STULTZ
is *Make:*'s digital fabrication editor. He is the founder of 3DPPVD, Ocean State Maker Mill, and HackPittsburgh.

It's in the Stars

Laser cut and light a model of the sky from a special night in your life

Written and photographed by Joe Spanier

JOE SPANIER is an engineer, father, and DIY CNC enthusiast. In his spare time he is president of his local makerspace, RiverCityLabs (rivercitylabs.space).

**Time Required:
A Weekend
Cost:
$100–$120**

MATERIALS

- » **RGB LED strip (6')** I used the kind that comes with a controller and remote
- » **Wire, 22ga** I used 4 colors, about 4' each
- » **Solder and flux**
- » **MDF, 2'×4' Sheet, ¾" thick** For internal spacer. 3 sheets of ¼" ply would work as well
- » **Birch ply, laserable, 2'×4', ¼" thick** You'll need 2½ sheets, just get 3
- » **Corner molding, wood, 1" leg** long enough to wrap around your full map frame
- » **Spray paint** I like Krylon or Rustoleum. Get primer, base color, and clear from the same brand.
- » **Glue, wood**
- » **Sandpaper** I used 220 with a palm sander
- » **Tape, painter's** ScotchBlue works great for how we are going to use it
- » **Screws, ½"** for the clips for the LED strip
- » **Brads**

TOOLS

- » Laser cutter and software
- » Wire cutters/strippers
- » Soldering iron
- » Helping hands
- » Drill and drill bits
- » Screwdriver
- » Miter saw or miter box
- » Cloth, lint-free
- » Alcohol, isopropyl

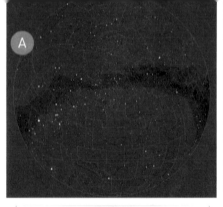

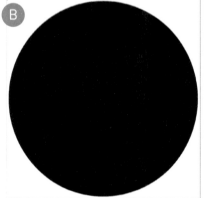

WITH MY ANNIVERSARY LOOMING, I NEEDED AN AMAZING PRESENT AND, BEING ME, I NEEDED TO MAKE IT MYSELF.
I got ahold of a friend to finish our quick fork of a star chart generator script we had started a while back. A few days later, I had made the best present for my wife that I've come up with yet: a glowing LED model of the sky from the night we got married. Here's how to make one too.

1. DESIGN THE STAR MAP

For the map, I used d3-celestials' interactive webform (armchairastronautics.blogspot. com/p/skymap.html), centering the sky at a specific date and time with a custom Python script to generate the star map and export an SVG. If you're feeling adventurous and have a decent know-how of Python, try making your own script to get a date-and-location-specific star array. Otherwise, use the SVG of the northern hemisphere sky I've uploaded to makezine.com/go/laser-star-map.

2. PROCESS THE SVG

The SVG that comes out of our export script (Figure Ⓐ) does not render well when opened in Inkscape (Figure Ⓑ), which is needed for many laser cutters. The different objects are broken into groups rather than layers, so we need to add them. Layers will help make different processes in the laser's software. I used four layers, named Grid, Stars, Constellations, and Milky Way.

3. BUILD THE BACK SHEET AND FRAME

First, cut a piece of ¼" ply to the final dimensions of your project for your back sheet. I made my frame 30"×22" — I wanted the map to be as large as possible so even the smallest star would get light though.

4. CUT THE SPACER

I used my CNC router to cut ¾" MDF into a spacer, to give the LEDs some distance to diffuse and be the right thickness to accommodate the LED controller. Any solid material will work, such as 3 sheets of laser ply laid up with wood glue. Be sure to allow slots for any wiring or IR receivers for the LEDs if yours has a remote.

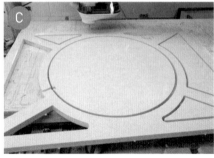

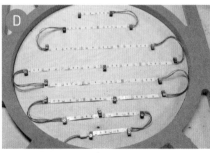

> **NOTE:** Make the inner diameter of the spacer slightly smaller than the outer diameter of the star circle (Figure Ⓒ). This will give support to glue the stars down and keep light from leaking out of the border.

5. DETERMINE LED PLACEMENT

Lay out your LED strips to go as close to full length as the cut lines will allow. Mine were spaced about 3" apart (Figure Ⓓ) and the lighting was extremely even and plenty bright. Experiment with your LEDs before gluing. Once you get them placed the way you want, mark them and solder jumpers to connect the strips — 3"–5" wires seemed to work well in mine.

6. CUT BACKGROUND SHEET

Find the nicest piece of your ply and cut it to the final size of your project. Then laser cut a circle the same size as your star map. Very lightly sand and spray paint (Figure Ⓔ) the background board and the edge trim. I like to prime everything that is bare wood, using 1–2 coats of sand-able primer from the same company that makes whatever base I'm using — that way I know the solvents will likely play nice together. Finish with 2–3 coats of clear.

7. MORE LASERSSSSSSS

I like to cover my project with blue painter's tape to keep smoke off of the light surface of the birch plywood. For my map, I ran each layer like this:
a. Cut Stars
b. Lightly vector-etch the astronomical grid
c. Slightly darker, etch the constellations
d. Milky Way — I broke this into 5 layers and power settings and raster etched each layer starting with the lightest layer and moving to the darkest.

I had to cut my map in 2 pieces because I wasn't able to find laserable birch ply with good inner plies in large enough sheets to fit the full map, so I split it along the upper line of the Milky Way (Figure Ⓕ).

Once cut (Figure Ⓖ on the following page), clean the tape off the project and clean any remaining bits off with a lint-free cloth and some isopropyl alcohol. Then take the clear coat you used on your background and put 2 coats on the map to seal it from moisture.

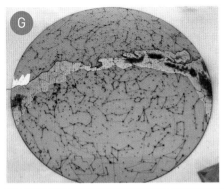

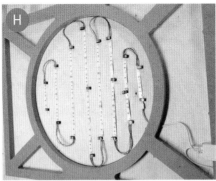

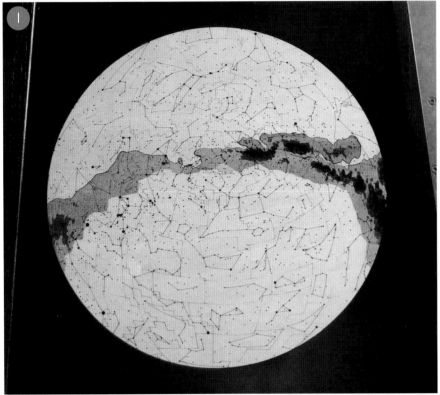

8. ASSEMBLE

Lay out the LEDs and route wires according to your marks on the back board. Screw them down with clips (I laser-cut thin tabs with screw holes for mine). Make sure to not tighten them down fully and pinch the strip — power them up and test before going forward. Put glue on the back of the inner brace and lay it down, again being careful to not pinch wires (Figure H). Place the LED controller (and IR receiver if using).

Put glue on the top of the inner brace and gently lay the background on the top. Shift it into position making sure the circle for the map is centered and the ledge for the map to rest on is even. Put a soft cloth on the dry, painted surfaces and add some weight to make sure everything is flat and compressed while the glue dries overnight.

Cut the trim to fit the edges of the frame. I used a few drops of glue and brads to hold the trim on. Once the trim is on look for any spots that need touch up on the paint. If your frame is thicker than your trim, lightly hit any exposed frame with base coat.

9. GLUE IN THE MAP

Put a small amount of glue on the rim of the brace. Gently place the map into the background and press it into the glue. Once it's set put another coat of clear over the whole project and let it dry (Figure I).

It's DONE!

Hang your map near a power outlet so you can light your LEDs easily. Be aware that this thing is probably pretty heavy so plan accordingly for hanging it. ◗

It's a Wrap

Make your print pop with graphics by dipping it in hydrographic film
Written by Shawn Grimes

HYDRO DIPPING IS A SIMPLE WAY TO APPLY GRAPHICS AND "SKINS" to your 3D prints. The process transfers images from a thin hydrographic film onto nearly any three-dimensional object by floating the image in a tub of water. Professionals have used this technique since the '80s for custom automotive and motorcycle parts, and it's recently become popular with printing enthusiasts.

The graphics come on clear film and can be purchased online. You apply it by carefully laying the film on top of the water and spraying it with an activator solution. This activator dissolves the film — the ink is left floating on the water's surface, ready to adhere to your 3D print. As you dip your 3D object into the water, the surface tension allows the graphic to curve around the contours of your 3D print.

PROCESS:

1. Make sure you are in a well-ventilated area and (optionally) give your 3D print a coat of spray-paint primer.

2. Cut a piece of film sized to cover your print (Figure A).

3. Depending on your print, you may want to glue a small handle on the bottom of it. A craft stick works great, but you can also use whatever scraps you have lying around, or a loop of tape.

4. Fill your bucket or tank with room temperature water. I applied strips of tape to help keep the design centered and off the sides of my container. Put on your gloves and place the hydrographic film on the water's surface with the dissolvable side down (Figure B). (Rub a scrap with wet fingers to test for this.)

5. Wait about 2–3 minutes. The graphic will wrinkle (Figure C) and then smooth back out as the film dissolves. You need this to happen before you spray the activator.

6. Shake the activator can vigorously and give the top of the graphic a light spray to make it sticky (Figure D). It's ready for application once it turns glossy — about 10–20 seconds (Figure E).

7. Hold your object by one end (or, optionally, by the handle you've affixed to it) and dip it slowly into the tank to allow the graphic to wrap around it (Figure F). Then give the object a twist to disconnect it from the floating ink.

8. Remove your print and allow it to dry overnight (Figure G).

Optional: Spray with a clear coat of the finish you desire (glossy or matte) to protect your print. ⊘

Time Required: 15–20 Minutes, 24 Hours to Dry
Cost: $40–$45

MATERIALS

» **Hydrographic film** infectedhydro.com and dip123.com are two online options for pre-printed graphics, or try making your own on inkjet-printable film from prostreetgraphix.com or similar
» **Spray activator**
» **Your 3D print**
» **Water** room temperature
» **Spray paint primer, white** (optional)
» **Spray paint, clear (optional)**

TOOLS

» **Bucket or tank** wide enough for your film to sit inside, and deep enough to fully dip your 3D print into
» **Gloves, latex or nitrile**
» **Scrap wood, craft stick, tape, etc. (optional)**
» **Glue gun and glue (optional)**

A

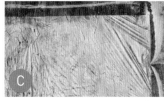
B

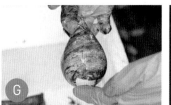
C

D

E

F

G

SHAWN GRIMES is the executive director at the Digital Harbor Foundation, using maker skills to develop creativity and productivity in youth and educators.

Hep Svadja

CNC
Step Stool

Cut this simplified version of a family heirloom to pass down to your future generations

Written by Matt Stultz

MATT STULTZ
is *Make:*'s digital fabrication editor. He is the founder of 3DPPVD, Ocean State Maker Mill, and Hack Pittsburgh.

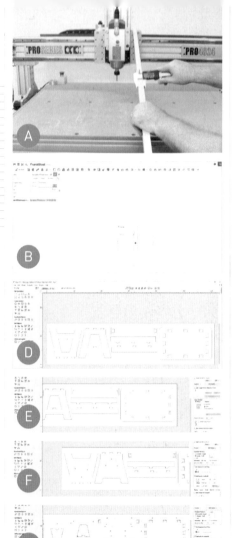

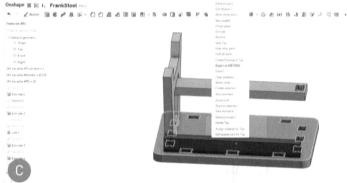

Time Required:
3–4 Hours
Cost:
$20–$50

MATERIALS
» **Wood, 44"×11" board at minimum,** larger if you need to split the parts up to fit a smaller CNC. I used poplar because it's inexpensive, cuts well, and looks great but you can use whatever wood you want.

TOOLS
» **CNC router** capable of cutting 380mm×230mm (~ 15"×9")
» **Router bit, straight flute cutting, ¼"**
» **CAM software** I use Vertric's VCarve. If you are a member of a hackerspace/makerspace, be sure to check out the hackerspace edition.
» **OnShape cloud-based CAD software,** free for makers
» **Chisel, small saw, or other sharp object**
» **Sandpaper**
» **Mallet**
» **Wood glue**
» **Eye and ear protection**
» **Small paintbrush (optional)** for gluing. A paper towel or your finger works as well.
» **Tung oil (optional)**
» **Paint or stain (optional)**

AS FAR BACK AS I CAN REMEMBER THERE WAS A FIXTURE IN ALL OF MY FAMILY MEMBER'S HOUSES: a little wooden stool. Purchased by my great uncle Frank, an engineer, from one of his co-workers for every female relative, my mother's copy was my breakfast table during Saturday morning cartoons; the fort for epic battles between my action figures; and my workbench for my first woodworking project around the age of 7. When my grandmother passed away, the one thing I asked for was her stool.

One day I spotted my wife sitting in front of our fireplace on my grandmother's stool and it hit me: I would redesign that stool to be cut on a router. I hope that my uncle Frank's family gift can spread to be the first workbench, drafting table, or thinking seat for a new generation of makers.

THE DESIGN
To make this version easy to cut, I squared things up a bit. The design is in OnShape (onshape.com) so you can easily modify it yourself for free. Get the file here: makezine.com/go/cnc-step-stool.

1. MEASURE YOUR MATERIALS
Measure the thickness of your material throughout the entire board for accuracy (Figure Ⓐ). Open the OnShape model, double click the "#Thickness" variable, input the measured thickness (Figure Ⓑ), and click the checkbox; it will adjust all the joints for press fit construction (although I do suggest using glue).

2. CREATE YOUR VECTORS
Right-click on a face of each part in the model and select "Export as DXF/DWG" (Figure Ⓒ) to create a 2D vector. You must click on the bottom surface of the stool top

and the cross brace to export the joint holes. Click the other two parts on the sides with the largest surface area.

3. SET UP THE JOB
Input your board's dimensions in VCarve (I cut in millimeters). Set your home origin to the bottom left of the part and the zero as the top of the board you are cutting. Import each DXF one at a time and use the Join tool to close up any vectors. Lay out your parts based on the size board you are using, leaving plenty of clearance on the corners. I was able to fit all the parts for the stool on a single 44"×11" board (Figure Ⓓ).

The stool is cut in three operations: a pocket (Figure Ⓔ), an inner cut (Figure Ⓕ), and an outer cut (Figure Ⓖ). A pocket removes all the material inside a selected boundary line, leaving a hole — select the Profile cut tool, and in the "Machine Vectors" section select "Pocket" for the dog-boned holes first. I like to use a ¼" two-flute straight bit. You only need to cut down 10mm deep for all the pockets in this design. If you want to go any deeper (or shallower), be sure to modify the "#PD" variable in the OnShape document before exporting it.

Next select "Inside" in the Profile cut tool to cut the inner sections of the legs. Vectors can be cut either inside the line, outside the line, or on the line. This cut should go all the way through your work piece so set the cut depth to the board's measured thickness. Before finalizing this operation be sure to add a few tabs to your part to prevent the piece from coming loose after it's cut and possibly damaging your job, your machine, or you.

Finally, setup the outside cut — this will be the majority of the parts. Choose "Outside" in the Profile cut tool and again add tabs. This job also needs to cut the

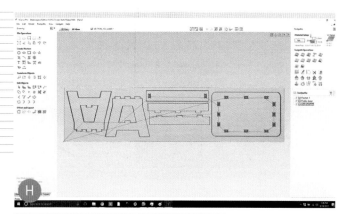

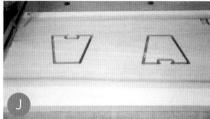

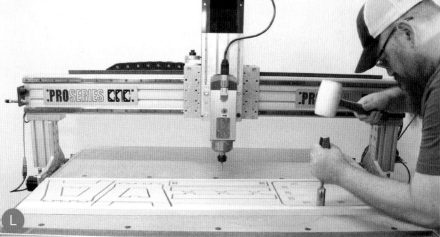

Matt Stultz, Mike Stultz

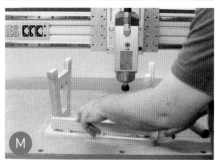

entire way through. When all your paths are set up (Figure H), export each job type as the G-code that matches up to your machine type.

4. GET CUTTING

Affix your material to the bed, and be sure to square it to the machine. Important: If you use screws to hold down your material, triple-check your paths to make sure the bit doesn't run into them and ruin your tooling. If you separated the G-code into individual files for each operation when you exported the files, start with the pocket job first.

SAFETY FIRST: Make sure you have proper eye and ear protection before running your CNC router.

After the pocket job (Figure I) is done, run the inner cut (Figure J) then the outer cuts (Figure K) to complete the milling.

5. ASSEMBLE

Remove the board from the bed and cut the tabs free (Figure L). Use sandpaper to clean up the edges. Add a little wood glue to each of the holes in the top of the stool, making sure it evenly coats the bottom of the hole and up the sidewalls. Now press the legs into place. You might need to use a mallet for this as the parts should fit tightly together, being careful not to damage the wood. Work from side to side on each leg until it is fully seated.

Put a little wood glue on the side of each leg coming up about an inch to help bond the sidewalls (Figure M) to the legs and stiffen up your stool a little. Again hammer the sides into the top, working from side to side until they are fully seated. Finally, add some wood glue to the holes in the cross brace and with the stool sitting right side up, slide the brace through the legs and down onto the pegs. Gently hammer if necessary (Figure N). Let the glue dry for a few hours.

6. FINISHING

Since it can see years of heavy use, finish the stool with some kind of protective coating. I like tung oil, which can be brushed on with a rag and worked into the wood. It will seal the surface of the wood helping it live up to the stresses a stool is put under. ●

Little Laser Show

Written by Evan Stanford

Create cool images with this 3D printed, hand-cranked device

EVAN STANFORD
is a software developer in Los Angeles, but he's always been interested in 3D modeling and mechanical engineering. 3D printers have allowed him to further explore these interests.

**Time Required:
2 Hours + lots of printing time
Cost:
$40**

MATERIALS
- » **3D printed parts** Download the files free at thingiverse.com/thing:2383299 and print them out or have a service do it for you.
- » **Laser pen**
- » **Rubber bands, light** Normal bands can provide too much force.

TOOLS
- » **3D printer and filament**
- » **Computer and project code** free download at github.com/EvanStanford/cams (only required if creating new patterns)

THE MECHANICAL LASER SHOW IS A HAND-POWERED DEVICE that quickly moves a laser to project an image. It's entirely 3D printed and all parts are open source and free to download. And it's easy to assemble by hand without tools.

HOW IT WORKS
The device moves the laser along a path repetitively, and thanks to the persistence of vision effect, you see the whole path at once. Turning the hand crank rotates 2 near-circular *cams*, and the laser acts as the *follower* that rides on top of both (Figure A). While rotating, each cam's radius increases and decreases to move the follower in the desired path; one cam is used for x-axis movement and one for y-axis movement. The cam profiles are precisely calculated for each target image. Every time the cams make a revolution, the laser traces the path once. The cams rotate about 5 times per second due to a 5:1 gear ratio between them and the hand crank.

CREATING THE CAM PROFILES
The cam profiles are calculated as a series of points that form a convex polygon. The input path (star, Batman, etc.) is also a series of points (Figure B). For each input point, a single point can be calculated on each output cam by doing a series of geometric transformations. I wrote an open

source Golang program to do the math. For more info, watch the video (youtube.com/watch?v=_dtBUiaAqRE) or check out the code (github.com/EvanStanford/cams).

THE DESIGN PROCESS
After I realized this device would be possible to create, I designed a complex solution. Then I went through several iterations to simplify and eliminate parts. My final design has 3 independently moving printed parts.

Then I looked at the parts' interactions. In my experience the best 3D printed fasteners are nuts and bolts, so the cams and crank are held in place by axle bolts. All axle bolts are threaded such that they tighten during use; in order to do this, I made one of the axles left-hand threaded.

Finally, I developed the cam profiles. I printed out several versions that were all a bit off before getting the math right. I modeled everything in SketchUp and wrote the code in Golang.

POSSIBLE IMPROVEMENTS
Currently the mechanical laser show can only draw continuous, connected paths. A simple fix would be to add a shutter wheel on one of the cams that could intermittently block light from the laser. To increase speed and accuracy, you could also convert the machine to use electric motor power instead of hand power. ⊘

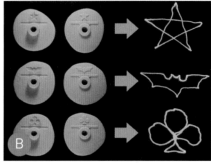

A

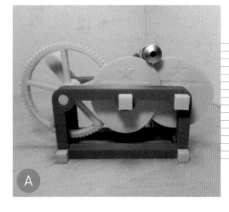

B

Evan Stanford

See the mechanical laser show in action and learn more at youtube.com/watch?v=_dtBUiaAqRE.

iPad Teleprompter

3D print your own visual text display Written by Seng-Poh Lee

Time Required:
30 minute assembly
+ lots of printing time
Cost:
$25

MATERIALS

- » **3D printed parts** download them at thingiverse.com/thing:1665711 You will print: camera risers (2), iPad arms (2), glass edge covers (3 — you will need to cut down two of them by 2"), platform base, camera base, tripod mount or Manfrotto mount, and glass corners (4)
- » **Machine screws and nuts, 6×32×¾" (4), 6×32×1" (12)**
- » **Machine screws and wing nuts 10×32×1¼" (2)**
- » **Machine screw, ¼×20×½"**
- » **Black fleece, ½ yard** or other dark material
- » **Binder clips of various sizes (6–8)**
- » **Glass or acrylic, 10"×8"**
- » **Blue painter's tape**

TOOLS

- » **3D printer and filament** or use a printing service
- » **2-part epoxy**
- » **Gorilla Glue (optional)**

I NEEDED A VIDEO CAMERA TELEPROMPTER RECENTLY and (unbelievably!) did not find one already on Thingiverse at the time, so I had to cobble one up quickly. This was designed for my specific needs, please modify it or design it to fit yours.

1. PRINT THE PARTS

Download the parts (Figure 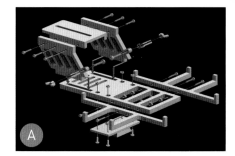) at thingiverse.com/thing:1665711. If you don't have a printer you can order them through a service such as Shapeways or Sculpteo.

2. ASSEMBLE THE PIECES

Screw the base platform to the iPad arms using six 6×32×1" machine screws and nuts, then insert the camera risers into the slots in the base and glue with 2-part epoxy. Screw the camera base onto the camera risers with the other six 6×32×1" machine screws and nuts.

3. ATTACH THE GLASS

Use the two 10×32×1¼" machine screws and wing nuts to attach the glass hinges to the camera risers. Neatly cover the edges of the glass on all sides with blue painter's tape. I tend to use two layers so that it gets firmly wedged into the hinges. Cut 2 of the glass edge covers down to 8 inches — I used a Dremel, but if you printed in PLA you can

use a craft knife. Place glass edge covers along three sides of the glass, and put all four corners in place. Carefully put the uncovered (but taped) glass edge into the hinges. If it feels loose add another layer of tape, and for even more stability put Gorilla Glue along the edge before inserting.

4. ADD THE BASE PLATFORM

Screw the base platform into the tripod mount (whichever style you chose) using the four 6×32×¾" machine screws and nuts. Pay close attention to which sides you place together — there are screw holes sunk so as to minimize the screw profiles; make sure the screw sinking holes are both on the outside. Screw the entire thing into your tripod, or insert it into your Manfrotto mount.

5. SET UP THE IPAD AND CAMERA

Put the iPad in the iPad arms, with your teleprompter app of choice loaded (Figure ). Attach your camera to the camera platform using the ¼×20×½" machine screw into your camera tripod mount. Try to balance your camera and iPad in such a way that the tripod mount is the center of gravity. PLA-printed stuff is not that strong, especially with the grain, and can easily break if too much weight is applied; you can adjust where the screw mount sits by sliding it along the track. Make sure to leave room

so that your lens doesn't hit the glass once it's raised (Figure ). Drape black fabric over the camera body, and attach the drape to the glass using binder clips.

Now just relax, and speak directly into the camera. ⊘

Hep Svadja

Toy Inventor's Notebook

HOLIDAY WINDOW STENCILS Invented and illustrated by Bob Knetzger

Time Required: 10–30 Minutes
Cost: $5–$25

MATERIALS
» **Cardstock or heavy paper**
» **Window Wax brand glass cleaner**
» **Acetate (optional)**

TOOLS
» **Computer and printer**
» **Hobby knife**
» **Sponge**
» **Tape (optional)**
» **Laser cutter or vinyl cutter (optional)**

HERE'S A FUN AND EASY WAY TO DECORATE YOUR WINDOWS FOR THE HOLIDAYS — and clean them at the same time, too! Cut out these holiday-themed stencils and use them to dab glass cleaner onto your windows: when it dries to a frosty white you'll have festive wintery windows! After the holidays, wipe off the haze to leave your glass clean and sparkly.

① PRINT AND CUT STENCILS
To make the stencils, download the designs from makezine.com/go/holiday-window-stencils and print them on cardstock or heavy paper. Use a hobby knife to carefully cut out the stencil shapes. For a longer lasting stencil, lay the design over acetate or other thin plastic sheet, then cut out. You can also download vector-art versions of the shapes that you can use with a laser cutter or vinyl cutter.

② APPLY DESIGNS TO WINDOW
Hold or tape the cut-out stencil onto your window, then use a sponge to dab the cleaner on, a little at a time, to fill the stencil. Done! (The original, extra smelly Glass Wax that I remember from years ago has now been replaced with a more environmentally friendly Window Wax formulation — whew!) ●

BOB KNETZGER is a designer/inventor/musician whose award-winning toys have been featured on *The Tonight Show, Nightline,* and *Good Morning America.* He is the author of *Make: Fun!,* available at makershed.com and fine bookstores everywhere.

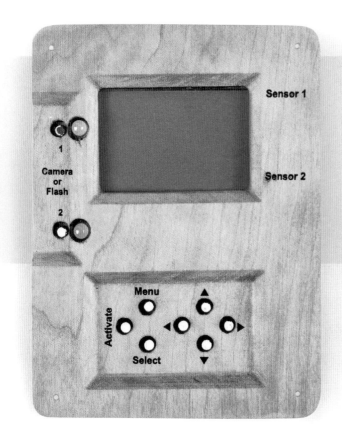

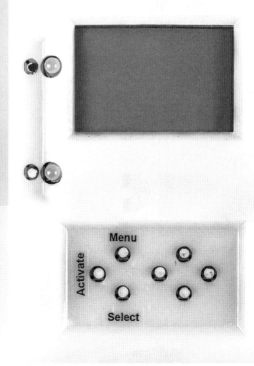

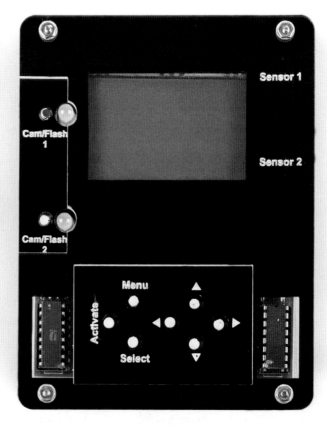

Make Your Case

Written by Art Krumsee

Prototyping adventures in CNC routing, SLA, FFF printing, and laser cutting

ART KRUMSEE has 30 years of experience as an IT director focused on internet and web development technologies. He works part-time as a consultant but increasingly he devotes time to everything from teaching meditation to creative projects at the Columbus Idea Foundry.

FOR BUILDING PROTOTYPES, WHAT ARE THE ADVANTAGES AND DISADVANTAGES OF 3D PRINTERS AND CNC ROUTERS?

I had no experience with 3D design or printing, but wanted to prototype a case for Maurice Ribble's brilliant Camera Axe kit. Mounted on an Arduino, the 4"×3.25" board included 9 switches, 2 LED indicator lights, and a small LED display. It was going to be a challenge — this version hadn't been designed to place in a box. I decided to compare results on two types of 3D printers (the LulzBot Taz 5, an FFF machine, and the Formlabs Form1+, an SLA machine), a ShopBot CNC router, as well as a 120W Trotec Speedy 400 laser cutter.

3D DESIGN

Learning CAD software was my biggest challenge. I settled on Autodesk's Fusion 360. It's enormously powerful, cloud-based, constantly improving, and free to use for hobbyists. I spent many hours watching YouTube tutorials and becoming comfortable with the software, which is on the level of learning Photoshop, before designing my case (Figure Ⓐ).

The workflow for the two 3D printers and the CNC router is similar:

1. Design the object in a CAD program and export in STL format.
2. Import the STL file into the tool's pre-processor.
3. Use the pre-processor to create supports, orient the design, and in the case of the CNC router, create the toolpaths.
4. Export the resulting configuration in G-code.
5. Using the machine's control software, load and execute the G-code.
6. When the piece is completed, finish as desired.

FROM DESIGN TO PRODUCTION AND BACK AGAIN

I started with the LulzBot. I found myself repeatedly printing, adjusting, and printing again. I set the print quality to low to get a usable result in about 4 hours. (The highest quality took about 8 hours.)

The LulzBot ships with Cura, its pre-processor. The default profiles produced good results. I printed the interior facing upward and automatically generated supports filling in the exterior recessed areas (Figure Ⓑ).

Supports were necessary, but they left irregularities that were difficult to sand away, particularly in recessed areas. After hours of sanding, I finished with a couple coats of spray acrylic.

My first prints bulged in the middle (Figure Ⓒ). Sanding this flat to mate with the other half was a significant headache. ABS plastic is durable but prone to warping. I switched to nGen from ColorFabb, which warped less and produced a consistent, stable, and (with finishing) attractive product.

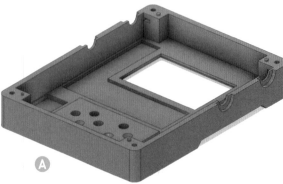

Ⓐ

Ⓑ

The lid's underside, showing infill lattice in yellow and supports of overhang areas in aqua.

Ⓒ

Hep Svadja, Art Krumsee

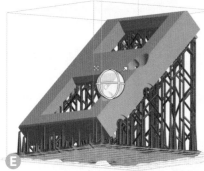

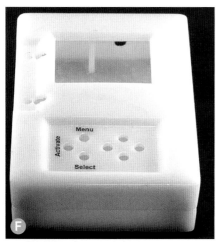

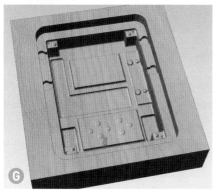

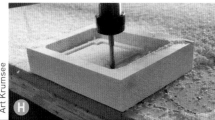

Filament printers typically print with a lattice infill rather than solid plastic. This saves time and material without a drastic loss of strength. But it made inserting screws more difficult, because it doesn't support threads very well. The solution was threaded heat-set brass inserts. Using a soldering iron, these slipped neatly into printed holes and melted solidly in place (Figure D).

WORKING WITH SLA

Much as LulzBot includes Cura, Formlabs includes PreForm with the Form 1+. The process of importing the component as an STL file, generating the G-code, and sending it to the printer is nearly identical (Figure E).

The heat-set inserts I'd used with the filament plastics don't work with SLA prints. Formlabs recommends creating pockets into which a nut will slide horizontally. The nut can be glued in place and then the screw is inserted through a hole from above. I was genuinely awed by the perfect quality of the holes and pockets in the final print, and of the print quality overall. The sides and edges of the case were sharp, clearly defined, and strong (Figure F).

The caveats are that SLA is expensive, slow, and messy (although the Form 2 uses cartridges which reduce some of the mess). One side of my case took about 9 hours at medium quality. The resin is sticky and can cause skin irritation, so gloves help. Parts come out coated with residual resin that must be removed, typically with several baths in alcohol. Finally, the prints are quite soft, and must be cured in UV light. Submerging the part in water for curing speeds the process.

CNC'ING ON THE SHOPBOT

I was concerned about the ability of wood to hold up to a router bit spinning at 12,000rpm so I doubled the thickness of the case walls to 6mm (Figure G).

After saving the CAD file in STL format,

It's astonishing that you can create a design once in a CAD program and use it to produce something using four very different processes.

I fired up VCarve Pro, the ShopBot pre-processor. Learning to use Cura and PreForm had been relatively simple. Not so with VCarve. One begins by defining the dimensions of the stock from which a component will be created. To hold costs down I used a 2"×6" cut into 6" sections.

The greatest challenge in VCarve is creating toolpaths. Once an object is imported, it can be broken into individual vectors, and then those can be used to define what the bits do. You must define a design in terms of specific tasks. For each task, one must choose an appropriate bit and then identify the route that bit will follow through the wood. Toolpath tasks include roughing out the interior (Figure H), fine tuning the interior, cutting out the external profile, drilling appropriate holes, etc. I used three router bits (¼" end mill, ⅛" end mill, ⅛" ball end) and two drill bits (⅛" and ¹⁄₁₆"). There are choices for each bit: how fast will it spin, how fast will it move, how much will one cut overlap with the previous one, etc. Fortunately, the defaults worked very well.

The movement of each bit is animated on the screen in VCarve Pro. I cut my first pieces in pine and was pleasantly surprised at their quality. When I moved on to a final print in cherry, I was amazed by the clean, well-defined shapes in a very complex object (Figure I). I realized I'd been excessively conservative in redesigning the wall thickness — in hardwood the results would have been fine with 3mm walls.

A FOURTH CONTENDER: LASER

After experimenting with CNC routers and 3D printers, round four began — I'd never used a laser cutter.

Rethinking the 3D design as a set of planar surfaces seemed restrictive. Acrylic sheet comes in defined thicknesses, ¹⁄₁₆", ⅛", etc. My original design had recesses in the top and the top's underside. Since the laser cutter can't create recesses, each of those levels would require its own piece

Art Krumsee

of acrylic (Figure **J**). But then too, the variance in height wasn't necessarily ¹⁄₁₆" or ⅛".

FROM FUSION 360 TO COREL

Going from a CAD design to the 120W Trotec Speedy 400 requires the additional step of putting the CAD design through a vector graphics program (I used CorelDraw). Settings for the laser cutter are similar to feeds and speeds for CNC.

> **NOTE:** *CorelDraw Graphics Suite accepts CAD drawings in DXF format. You have to turn off "Capture Design History" in Fusion 360 to export a DXF file from a sketch. You can turn it back on afterward, but you'll have lost all previous design history.*

LIMITATIONS AND WORKAROUNDS

The middle section of my design stood about 1" tall. However, when I tried to cut 1" acrylic with the Trotec, it went slowly and the result was sad, so I split each endcap into two pieces horizontally and cut them from ½" acrylic. The sections held their shape, but lost perhaps ¹⁄₁₆" of height. I used some shorter standoffs to mount the board and that solved the height problem.

The rear endcap has a U-shaped opening for a power cable (Figure **K**), which required that it be rotated so that the side could be cut. I set an acrylic block in the upper left corner of the Trotec and set a "Marker" on the Job Control software in order to give a replicable position.

SCREWS AND LABELS

I adjusted the settings to engrave very deeply into the ⅛" acrylic for the base. By engraving halfway through the acrylic, I could create counterbore holes that nicely held the 2-56 nuts.

Unlike 3D printers and CNC routers, creating labels with the laser engraver is a piece of cake. Vector graphics editors are adept at positioning text. I edited the design for my "lower lid" to add the labels and

printed it on black acrylic (Figure **L**). With the masking paper still in place, I filled the letters with white paint. Then I cut the upper lid in clear acrylic to protect the lettering (Figure **M**).

CONCLUSIONS

It's astonishing that you can create a design once in a CAD program and use it to produce something using four very different processes.

From setup to finishing, the ShopBot took 8½ hours, the LulzBot Taz 5 took 9½ hours, and the Form 1+ took 13 hours. The parts took less than an hour on the laser cutter, and assembly only took about 15 minutes.

Materials do matter. While plastic is modern, wood has some great characteristics like impact resistance and remarkable strength. Each can be painted any color of the rainbow.

Then there's the fun factor. While the Form1+ produced quality and accurate parts, the process of handling resin and curing the parts is, to use the technical term, "icky." The ShopBot is great fun, but I wasn't comfortable leaving it unattended. As for the filament printer, it's nice to start a print, monitor it off and on to completion, and go about other tasks. Laser-cut parts look finished, don't require sanding, and they're quick — the jobs took minutes, sometimes seconds. The machine felt exceptionally well designed with easy to use software, even for beginners. Of the technologies I've tried, laser cutting is the first I'd recommend to someone for a quick, easy build.

Personally though, I bought a filament printer, and I'll do my final prints on an SLA printer. It produces beautiful results and minimizes time spent sanding. A bit of touch-up sanding and a couple coats of UV-resistant acrylic (to prevent the resin becoming brittle) and I was on my way. Having said that, the case printed in cherry is the one I'm most proud of. I do love wood. Call me a prototyper with an old-fashioned sense of beauty! ◒

> *Of the technologies I've tried, laser cutting is the first I'd recommend to someone for a quick, easy build.*

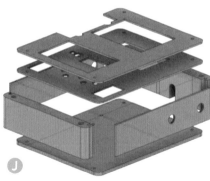

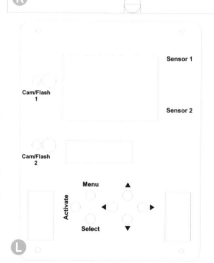

Skill Builder *Digital Joinery Design*

The Perfect Fit

Apply digital joinery techniques for better parts assembly

Written by Tasker Smith

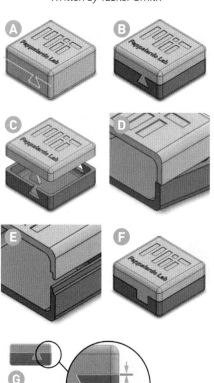

Class of printer	Professional	Hobbyist
3D printer tested	uPrint	MakerBot Mini
Material	ABS Plus	PLA
Gap tolerance	.005"/.127mm	.015"/.381mm

Gap tolerance: adjusted for 3D printing to achieve the desired type of fit

TASKER SMITH is a technical instructor at the Massachusetts Institute of Technology. He mentors students in the practical use of digital fabrication technologies and the process of iterative prototype development.

JOINERY IS A CENTURIES-OLD WOODWORKING TECHNIQUE that remedies the problem of seam separation between assembled components. Fitted joints have many advantages, like precise alignment of components and the ability to connect exclusively by the force of friction. We can apply this thinking to digital design and drastically improve the process of assembling multipart models.

The great advantage of digitally designed and fabricated joints is that special layouts, tooling, and woodworking fixtures aren't required. If configured properly, a 3D printer, laser cutter, waterjet, or CNC router will produce repeatable features that press together with remarkable consistency.

CREATING JOINT DETAILS

1. Project a parting line/joint detail on an object (Figure **A**).
2. Next, break the object into two segments by extruding a cut through the object (Figure **B**).
3. Save the individual segments as separate parts (Figure **C**)

BENEFITS OF EXTRUDED PARTING LINES

» **Simplicity** — If adjustments are required, they can be performed on a single parting line feature.
» **Alignment** — Aligning two 3D printed parts can be challenging. This is particularly true when bonding components together with quick curing adhesives like cyanoacrylate. Adding mating features takes the guesswork out of alignment and ensures repeatable location of components time after time.
» **Unambiguous Orientation** — Symmetrical designs often don't have obvious orientations. Adding asymmetrical alignment features eliminates this ambiguity, and can actually make it impossible to assemble segments improperly.
» **Mechanical fastening** — Dovetails (and other parting line details with undercuts) can hold components together without adhesives or external fasteners. This is particularly useful when components need to be assembled and disassembled.
» **Adhesion** — Permanent adhesion

of components requires good part preparation and selecting the proper adhesive, and is also influenced by the amount of surface area being bonded. A contoured parting line increases contact area — which results in a stronger bond.

JOINERY DESIGN GUIDELINES

Keep it simple and remember these fundamentals of joinery:
» **Butt Joints** are easy to design, but difficult to align (Figure **D**).
» **Lap Joints** align nicely, but can be fussy and fragile when 3D printed (Figure **E**).
» **Dovetail joints** lock features together and can be assembled and disassembled easily, as shown in Figure B.
» **Square keys** align features precisely and are well suited for gluing.
» **Asymmetrical location of keys** results in unambiguous orientation when assembling components (Figure **F**).

3D PRINTING GUIDELINES

Tolerances should always be considered when designing components intended to fit together. Dimensions of 3D printed parts will vary from the digital source files, and this discrepancy must be accounted for. Tight tolerances work with professional 3D printers, but hobbyist systems require a looser fit. We experimented both with a Stratasys uPrint and a MakerBot Mini, and found that the accuracy of the uPrint required minimal clearance (.005 in/ .127mm) to create a sliding fit between components, while the MakerBot required additional clearance (.015 in/.381mm) to achieve a similar fit (Figure **G**). There are many factors to consider (machine type and class, material, component orientation, layer thickness, etc.), so it's essential to test.

PROCEDURAL BEST PRACTICES

1. Isolate mating features and print small sections to test the tolerances of your machine — anything that's too tight or too loose can be worked out with a minimal investment of time and material.
2. Print test parts with the same printer settings and in the same orientation as final parts to get repeatable results. ◙

Tasker Smith

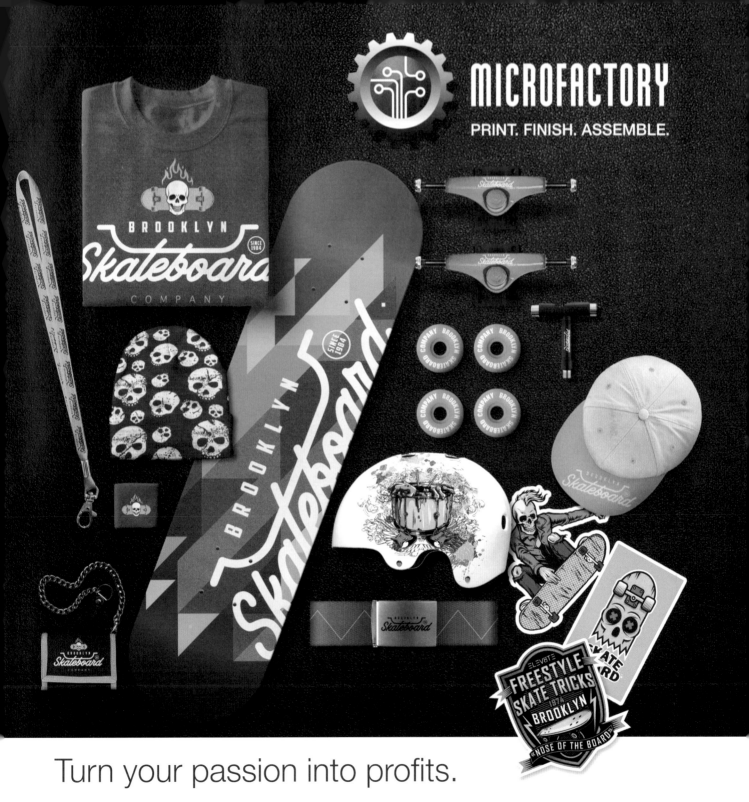

Skill Builder *Mold Making for Metallics*

Cold Casting
Techniques

Skip the metalworking and use this process for authentic looking metallic figures

Written and photographed by Evan Morgan

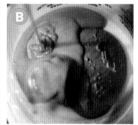
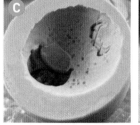
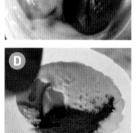

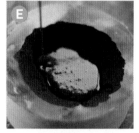
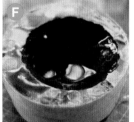
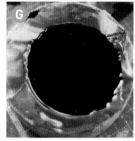
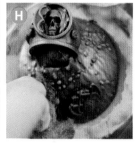

LET'S CREATE A REAL METAL RESIN PIECE USING THE COLD CAST METAL METHOD. Once polished, this gives a realistic finish due to the real metal powder that's mixed in.

MAKE THE MOLD

I recommended you choose a model that has a flat base, so you can use a one-part "dump mold." This yields a very clean surface finish with no seam. **Super-glue** your model to a flat surface (like tile or wood) and put a **disposable plastic cup** over the top, with ½" of clearance all around. Apply **hot glue** around the bottom to create a seal (Figure **A**).

Don your **respirator** and **nitrile gloves**. Mix a batch of **silicone rubber** according to the manufacturer's instructions. Cut off the bottom of the cup, and pour in the silicone from a high height to prevent bubbles (Figure **B**). Pour until it covers the model, and leave it to cure.

Once fully cured, remove the cup and demold your model by peeling at the edges until you can pull it away freely. Your mold is ready (Figure **C**).

CAST AWAY

Pour a little **metal powder** (copper, bronze, etc.) into the mold (Figure **D**) and **brush** it around. This helps create a better finish. Tap out any excess.

Most **rotocasting resin** (such as Smooth-Cast 65D) comes in two containers, A and B, requiring you to mix equal parts of each to activate it. Measure just enough to coat the surface of your mold. Add a scoop of the metal powder to part B of the resin and mix thoroughly. Then mix parts A and B together in a separate cup and pour into the mold (Figure **E**).

Slowly rotate and angle the mold so that the resin covers the entire surface area. Keep going until the resin has gelled and can no longer move. This should result in an even coating (Figure **F**).

Measure out a second batch of resin, enough to fill the mold completely. Add a few drops of **resin pigment** (to closely match your metal color) to part B before mixing.

Pour the pigmented resin into the mold slowly until filled to the top and allow to cure thoroughly (Figure **G**).

FINISH UP

Once fully cured, carefully peel each edge of the silicone until you can easily pull the cast part out.

To bring out the metal finish, buff the surface with **0000 grade steel wool** (Figure **H**) or a **polishing wheel**. Lightly buff back and forth until the desired surface sheen is achieved.

GOING FURTHER

For a finer finish, try high-gloss **polishing compound**, or oxidization to give the metal an aged look. ◐

EVAN MORGAN is a full-time self-employed model maker specializing in mold making and resin casting. Find him on YouTube at youtube.com/model3devan and on Instagram @evanmorgan93.

Make: *It Real*

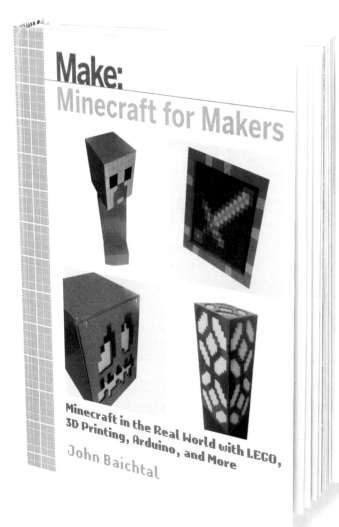

MINECRAFT FOR MAKERS
By John Baichtal $20

Explore the intersection of this beloved computer game with reality by taking familiar objects from the *Minecraft* world and physically constructing them in real life.

MODERN LEATHERWORK FOR MAKERS
By Tim Deagan $30

Take leather crafting into the 21st century with this complete guide that marries traditional skills to the latest CNC and 3D printing technologies.

GETTING STARTED WITH THE MICRO:BIT
By Wolfram Donat $20

The micro:bit, a tiny computer provided to students all over the U.K., is now available for anyone to purchase and play with! Learn to master it and get started working with IoT.

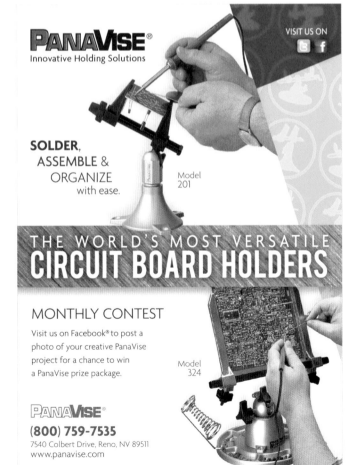

SHOW&TELL

Dazzling projects from inventive makers like you
Prank Cup Edition

Sharing what you've made is half the joy of making! To be featured here, show us your photos by tagging #makemagazine.

Written by Matt Stultz

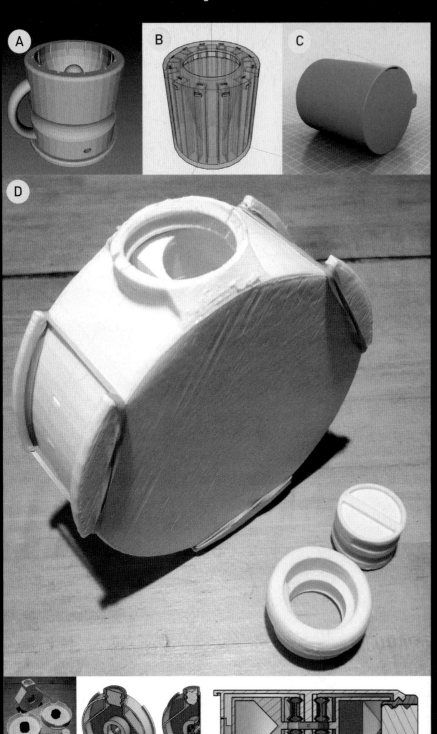

We had some really great entries for our Worst Cup Ever challenge. Most designs fell into one of three main categories: siphons, straws, and handedness. While there were lots of takes on these and extra tricks, the following were our favorites.

Ⓐ SIPHONS

Inspired by the classic Pythagorean cup, these cups use a system of internal chambers to create a suction that pulls liquid out of the cup and onto the drinker's lap when they attempt to drink. The example shown here is the well-thought-out Twisted Tantalus Cup by **Andrew Moore**.

Ⓑ STRAWS

Quite a few entries made it nearly impossible for you to access the liquids at all, if not for their hidden straws to sip from. Some of our favorites used this method, like this Tricky Tumbler by **Jonathan Tindal**.

Ⓒ HANDEDNESS

These designs took a slight twist on the Pythagorean cup, which relies on the victim being right handed and using the cup with their dominant hand, which then triggers the siphon. If you simply turn the cup 180° then you can drink safely. I especially loved the "April Fool's Cup" designed by **Calvin Iba** because it does such a great job of looking like a normal mug that it's bound to trick the unprepared.

Ⓓ THE WINNER

While not strictly a cup, there was one entry that really stood out for taking the time to come up with a creative solution and really putting in the engineering work to pull it off.

The Canteen of Denial has an internal set of bearings and a top-heavy design. When tilted up, the internal chamber spins to keep the contents from being consumed. The canteen has a secret cap with a center plug. If just the plug is removed, the rest of the cap locks the internal chamber in place, allowing the owner to sip the contents.

We loved the thought and effort put into designing this object, and were happy to announce **Kiefer Read** as the winner — and we can't wait to see what he'll make with sponsor SeeMeCNC's new H2 Delta Printer! ⊘